符号中国 SIGNS OF CHINA

U0483676

符号中国 SIGNS OF CHINA

符号中国 SIGNS OF CHINA

中国园林

THE CHINESE GARDEN

"符号中国"编写组◎编著

图书在版编目(CIP)数据

中国园林：汉文、英文/"符号中国"编写组编著. —北京：
中央民族大学出版社，2024.3
（符号中国）
ISBN 978-7-5660-2282-0

Ⅰ.①中… Ⅱ.①符… Ⅲ.①园林艺术—介绍—中国—汉、英 Ⅳ.①TU986.62

中国国家版本馆CIP数据核字（2024）第016751号

符号中国：中国园林 THE CHINESE GARDEN

编　　著	"符号中国"编写组
策划编辑	沙　平
责任编辑	杨爱新
英文指导	李瑞清
英文编辑	邱　械
美术编辑	曹　娜　郑亚超　洪　涛
出版发行	中央民族大学出版社
	北京市海淀区中关村南大街27号　　邮编：100081
	电话：（010）68472815（发行部）　传真：（010）68933757（发行部）
	（010）68932218（总编室）　　　　（010）68932447（办公室）
经 销 者	全国各地新华书店
印 刷 厂	北京兴星伟业印刷有限公司
开　　本	787 mm×1092 mm　1/16　印张：10.625
字　　数	138千字
版　　次	2024年3月第1版　2024年3月第1次印刷
书　　号	ISBN 978-7-5660-2282-0
定　　价	58.00元

版权所有　侵权必究

"符号中国"丛书编委会

唐兰东　巴哈提　杨国华　孟靖朝　赵秀琴

本册编写者

吕明伟

前言 Preface

　　中国园林从三千多年前的商周时期发展至今，已经逐渐形成了独特的风格，在世界造园艺术中独树一帜。

　　和欧洲园林理性、规整的艺术风格截然不同，中国园林最大的特色就是将自然山水的景色，浓缩集中到园林的景致之中，无论是空间布局的设计，还是借景、对景、框景、障景等造景手法的运用，中国园林都遵循着"虽由人作，宛自天开"的原则，对地形地貌进行少量的加工和改造，使人们亲近自然的愿望得到最大限度的满足。园林中的一山一水、一花一草、一楼一

The history of Chinese gardens dated back to the Shang and Zhou dynasties over 3,000 years ago. Since then, Chinese gardens have gradually developed a unique style and established a new school in the world of landscape gardening art.

　　Entirely different from European gardens — rational and regular in artistic style, the biggest feature of Chinese gardens is that it combines natural scenery with miniature gardens, which is smartly reflected in its spatial layout, and the application of landscape gardening techniques such as borrowed scenery, oppositive scenery, enframed scenery and obstructive scenery. Following the principle that the works of men should match that

亭、一桥一塔，在设计者精巧的规划之下，组合成许许多多令人惊叹的画面和意境。

为了使读者对中国园林有进一步的了解，本书就中国园林的类型、造园艺术和园林建筑等方面，结合精美的实景照片与手绘图片，对中国园林进行系统而深入的介绍，希望读者通过本书感受中国园林之美的同时，还能够领悟到中国传统文化的博大精深。

of heaven, the terrain and landform of a garden is just slightly changed and improved, which highly satisfies people's desire to be close to nature. Every single hill, stream, flower, grass, building, pavilion, bridge and pagoda is smartly planned and arranged by the designer, forming many remarkably impressive pictures and images.

To help readers further understand Chinese gardens, this book intends to give a systematic but understandable introduction to Chinese gardens from aspects of types, landscape gardening and the architecture of gardens, and provide exquisite photographs and hand-painted pictures. It is hoped that readers can comprehend the profound traditional Chinese culture when appreciating the beauty of Chinese gardens.

目 录 Contents

中国的园林
Gardens in China 001

皇家园林
Royal Gardens 002

私家园林
Private Gardens 018

自然景观园林
Natural Landscape Gardens........................... 033

中国园林的造园艺术
Gardening Art of the Chinese Gardens............ 047

空间布局
Spatial Layout 048

造景手法
Gardening Techniques.................................... 051

山石的布置
The Arrangement of Rocks............................. 061

水景的设计
Design of Water Scenery 075

花木的点缀
Ornament of Plants 085

中国园林中的建筑
Architectures in Chinese Gardens 093

园门建筑
Garden Gates.. 094

宫殿建筑
Palaces 100

厅堂建筑
Parlors & Halls 105

楼阁建筑
Storied Buildings & Towers 110

轩馆斋室建筑
Xuan, *Guan*, *Zhai* and *Shi* 113

游赏建筑
Landscape Constructions 119

园林建筑装饰
Decorations of the Architectures.................... 137

中国的园林
Gardens in China

  中国的园林虽然数量众多，形式各异，但基本上可以分为皇家园林、私家园林和自然景观园林。

Although there are numerous and various gardens in China, basically, they can be classified into royal gardens, private gardens and natural landscape gardens.

> 皇家园林

皇家园林，就是封建社会仅供历代帝王和皇室成员游赏的园林。皇家园林大多规模宏伟、占地辽阔、布局完整、功能齐全，园中建筑近似皇家宫殿建筑，辉煌而华丽。园林布局一般强调规则、均衡、对称等造园手法，以显皇家的威严与气派。

最早的皇家园林出现在先秦时期，早期主要是帝王打猎游乐的场所，所以占地面积广大，以自然景观和豢养动物为主，并不以建筑为主，称为"苑囿"。唐宋皇家园林主要集中在长安（今西安）、洛阳、东京（今开封）、临安（今杭州）等地，数量多且规模宏大，显示出泱泱大国的气概。现存的皇家园林多为明清时期构建，主要分布在北京、河北两地。

> Royal Gardens

Royal gardens are gardens exclusive to ancient emperors and royal families in feudal society. Most of them are of large scales and complete layouts, which cover a vast expanse of land and are well functioned. Buildings in the garden are similar to royal palaces, which are brilliant and gorgeous. Generally, the layout of the garden emphasizes landscape gardening techniques like rules, balance and symmetry, so as to showcase the royal dignity and authority.

The earliest royal garden appeared before the Qin Dynasty. At the early stage, royal gardens were called *Yuanyou* in Chinese, which were generally imperial hunting parks with natural landscapes instead of buildings. Animals were kept in the gardens for emperors' hunting tour. Royal gardens built in the Tang and Song dynasties were mainly located in Chang'an

皇家园林可分为皇城御苑、离宫御苑和行宫御苑。皇城御苑一般建在京城里面，与皇宫毗连，相当于私家宅院；离宫御苑多建在郊外风景优美、环境幽静的地方，是京城皇宫之外，皇帝长期居住并处理朝政的地方；而行宫御苑则供皇帝游憩或短期居住之用。

(present Xi'an City), Luoyang, Dongjing (present Kaifeng City) and Lin'an(present Hangzhou City). Royal gardens in this period were of large numbers and scales, which manifested the spirit of a great nation. Existing royal gardens are mainly located in Capital Beijing and Hebei Province, and most of them were built during the Ming and Qing dynasties.

Royal gardens can be classified into *Huangcheng*(capital) royal gardens, *Ligong* royal gardens and *Xinggong* royal gardens. *Huangcheng* royal gardens were generally built inside the capital, next to the palace, like a courtyard in a private house; *Ligong* royal gardens, mostly built in the beautiful and tranquil countryside outside the imperial palace in the capital, were the place where emperors usually lived and worked. And *Xinggong* royal gardens were built for the short-term tour and stays of emperors.

● 西汉建章宫复原图

建章宫是西汉武帝刘彻于太初元年（前104年）建造的宫苑。在建章宫的西北部，开有大池，名叫太液池，这是一片以园林为主的区域。池中堆筑了三个岛屿，象征传说中的瀛洲、蓬莱、方丈三座仙山。

Palinspastic Map of *Jianzhang* Palace of the Western Han Dynasty (206 B.C.-25 A.D.)

Jianzhang Palace was a royal garden built by Liu Che, Emperor Wu of the Western Han Dynasty, in the first year of Taichu (104 B.C.). In the northwest of *Jianzhang* Palace, there was a big pool named *Taiye* Lake (Lake of Grand Liquid). The area of *Taiye* Lake emphasized gardens. In the lake there were three islets which symbolized the three legendary mountains where the immortals lived: *Yingzhou*, *Penglai* and *Fangzhang*.

三山五园

　　三山五园是清朝皇家行宫苑囿的总称。三山指万寿山、香山和玉泉山；五园包括香山静宜园、玉泉山静明园、万寿山清漪园、圆明园、畅春园五座大型皇家园林。"三山五园"始建于清康熙时期，兴盛于乾隆时期，代表着中国皇家造园艺术的精华。在全盛时期，自海淀镇至香山，分布着90多处皇家离宫御苑与赐园，这些园林有人工山水园、天然山水园，也有天然山地园，连绵二十余里，蔚为壮观。大多数园林在1856—1860年的第二次鸦片战争中被焚毁。

The Three Hills and Five Gardens

The Three Hills and Five Gardens refer to all the royal gardens of the Qing Dynasty (1616-1911). The Three Hills are the Longevity Hill, the Fragrant Hill and the *Yuquan* Hill (Jade Spring Hill). The Five Gardens include five large-size royal gardens: *Jingyi* Garden on the Fragrant Hill, *Jingming* Garden on the *Yuquan* Hill, *Qingyi* Garden (Garden of Clear Ripples) on the Longevity Hill, *Yuanming* Garden (the Old Summer Palace) and *Changchun* Garden. The Three Hills and Five Gardens were first established in the Kangxi period and flourished during the Qianlong period, representing the essence of the Chinese imperial landscape gardening art. During its prime time, there were over 90 royal gardens located from Haidian Town to the Fragrant Hill. These gardens magnificently stretch more than 20 *li* (over 6 miles), among which were artificial landscape gardens, natural landscape gardens and natural hill gardens. However, most of them were burned to the ground during the Second Opium War (1856-1860).

- 《三山五园图》局部（清）
  Part of *The Three Hills and Five Gardens* (Qing Dynasty, 1616-1911)

故宫御花园

御苑，是帝王宫廷的延伸，在布局上一般有鲜明的轴线，建筑物之间也有相互对称的特点。如今保存完好的御苑有北京故宫内的御花园、建福宫花园和慈宁宫花园。

御花园位于故宫中轴线的最北端，在坤宁宫后方，明代称为"宫后苑"，清代称"御花园"，始建于明永乐十八年(1420年)。

御花园占地约1.1万平方米，全园南北宽80米，东西长约140米，有建筑20余处。园内建筑以钦安殿为中心，布局对称而不呆板，舒展而不零散，玲珑别致，疏密合度。园内现存古树160余株，散布各处，青翠的松、柏、竹间点缀着山石，形成了四季长青的园林景观。钦安殿左右有4座亭子：北边的浮碧亭和澄瑞亭，南边的万春亭和千秋亭，造型纤巧秀丽，为御花园增色不少。

倚北宫墙用太湖石叠筑的石山"堆秀"，山势险峻，磴道陡峭，叠石手法甚为新颖。山上的御景亭是皇帝和皇后重阳节登高的去处。园中奇石罗布，佳木葱茏，其中的古柏藤萝都有数百年的历史，将花

the Imperial Garden in the Forbidden City

Royal Gardens are the extension of the imperial palace. Generally they have distinct axes on the layout, and the buildings are in symmetry. Today, the Imperial Garden, *Jianfu* Palace Garden and *Cining* Palace Garden in Beijing Forbidden City are Royal Gardens that are still well-preserved.

The Imperial Garden is at the northern end of the central axis of the Forbidden City and at the rear of *Kunning* Palace (Palace of Earthly Tranquility). First built in 1420, it was called Palace Courtyard in the Ming Dynasty (1368-1644), and the Imperial Garden in the Qing Dynasty (1616-1911).

The Imperial Garden covers approximately 11,000 square meters, stretching 80 meters from north to south and around 140 meters from east to west. More than 20 buildings in the garden take *Qin'an* Hall as the center and the layout is symmetrical but not rigid, loose but not disordered. The whole garden looks elegant and well-ordered. More than 100 old trees exist in every corner of the garden. Decorated with hills and rocks, these green pine trees, cypress trees and bamboos form an evergreen landscape. There are two

• 养性斋

养性斋位于御花园西南角，是一座两层楼阁式的书房，是清代帝王读书、休息的地方。

Yangxing Zhai (Lodge of Spiritual Cultivation)

Yangxing Zhai is located at the southwest corner of the Imperial Garden. It is a two-storey study where emperors of the Qing Dynasty (1616-1911) studied and relaxed themselves.

• 千秋亭

千秋亭是一座三间四柱的方亭，顶部两层檐，上圆下方。整个亭子装饰富丽而富有生气。

Qianqiu Ting (Pavilion of One Thousand Autums)

The *Qianqiu Ting* is a square pavilion with three rooms and four columns. On the top there are two layers of eaves with the top one round and the other square. The whole pavilion is luxuriously decorated and appears to be full of vitality.

- 海参石盆景

海参石盆景长近80厘米，高60多厘米，造型犹如一个个小海参挤在一处，尤为真实生动。

Penjing (Bonsai)—Sea Cucumbers Stones:
The *Penjing* of Sea Cucumbers Stones is nearly 80 cm long, over 60cm high. The overall shape looks like many sea cucumbers huddling together, which is very vivid.

- 绛雪轩

绛雪轩位于御花园东南角，与西南角的养性斋相对。平面呈"凸"字形，与养性斋的"凹"字形平面也相对应。

Jiangxue Xuan (Bower of Crimson Snow):
Jiangxue Xuan is located at the southeast corner of the Imperial Garden, faced with *Yangxing Zhai* in the Northwest. It is in a convex shape, echoing with the concave shape of *Yangxing Zhai*.

- 故宫御花园示意图

Sketch Map of the Imperial Garden in the Forbidden City

园点缀得情趣盎然。园内甬路均以不同颜色的卵石精心铺砌而成，组成900余幅不同的图案，有人物、花卉、景物、戏剧、典故等，沿路观赏，妙趣无穷。

- **木变石盆景**

  木变石盆景置于绛雪轩前，虽为石质，却看似一段朽木，高达130厘米，令人称奇。

  Penjing (Bonsai)—Wood Alexandrite

  The *Penjing* of Wood alexandrite is placed in front of *Jiangxue Xuan*. Though a piece of stone, the 130cm high *penjing* looks like a piece of rotten wood which is slender and amazing.

pavilions on each side of the *Qin'an* Hall: *Fubi* Pavilion and *Chengrui* Pavilion in the northern part, and *Wanchun* Pavilion and *Qianqiu* Pavilion in the southern part. These pavilions are in a delicate and beautiful style, which enrich the beauty of the Imperial Garden.

Near the north wall of the Palace is the *Duixiu* Hill (Gathering Beauty Hill), a little artificial mountain with *Taihu* stones. The small mountain is very steep, and the way that the stones piled is very novel. *Yujing* Pavilion sits on top of the mountain and the emperor and his wife would climb up to the pavilion on the Double Ninth Festival. Rare stones and tall trees can be seen everywhere in the garden: old cypress trees and wisteria of several thousand of years make the garden full of enjoyment. Paths in the garden are well paved with pebbles of different colors, and composed more than 900 different pictures such as pictures of human figures, flowers, scenery, dramas and stories, etc. Walking along the paths, one can enjoy all the beauty around and feel inexhaustibly delighted.

颐和园

　　离宫御苑与内廷宫苑的规模相比要大得多，通常建在京城附近风景优美的山林郊外。位于北京西北郊的颐和园，是现存规模最大、保存最完好的离宫御苑。

　　颐和园距北京城区15千米，占地约293公顷，始建于清乾隆十五年（1750年），历经15年竣工，原名"清漪园"。咸丰十年（1860年），清漪园被英法联军焚毁。光

• 俯瞰颐和园
　A Bird's-eye View of the Summer Palace

The Summer Palace

Compared with the inner court, the *Ligong* Royal Garden is much larger in size. It was often built in the scenic outskirts of capital. The Summer Palace, located in the northwest of Beijing, is the largest existing and best-preserved *Ligong* Royal Garden so far.

　　The Summer Palace is 15 km away from the downtown of Beijing, and covers an expanse of 293 hectares. It was begun in 1750 (the 15th Reign Year of Emperor Qianlong of the Qing Dynasty) and completed 15 years later. Originally it was named *Qingyi* Garden (Garden of Clear Ripples). In 1860 (the 10th Reign Year of Emperor Xianfeng of the Qing Dynasty), it was burned to the ground by the joint Anglo-French Forces. In the year of 1888, Empress Dowager Cixi appropriated 30 million taels of silver into the reconstruction of the garden and renamed it the Summer Palace.

　　The Summer Palace manifests extraordinary landscape gardening techniques. Natural landscapes were smartly adopted in the construction

绪十四年（1888年），慈禧太后动用3000万两白银，将其重建，改称"颐和园"。

颐和园的造园技艺高超，巧借天然山水，体现自然之趣，高度表现了中国皇家园林壮丽、恢宏的气势。颐和园以昆明湖、万寿山为整座园林的基架，以杭州西湖美景为参照，结合江南私家园林的造园手法，兼具皇家园林的富丽和江南私家园林的精巧，是中国保存最完整、规模最大的皇家园林之一。

of the garden. The garden reflects natural interest and highly represents the magnificent and impressive manner of Chinese royal gardens. The Summer Palace combines the luxury of royal gardens and the delicacy of private gardens, showing them in one single garden: It takes *Kunming* Lake and the Longevity Hill as the basic framework and the beauty of Hangzhou West Lake as a reference, combining the landscape gardening techniques of private gardens in the southern areas. Today, the Summer Palace is one of the largest and the most intact royal gardens preserved in China.

The Summer Palace landscapes can be divided into three regions: political activity area (represented by *Renshou* Hall), living area (represented by *Leshou* Hall, *Yulan* Hall and *Yiyun* House), scenic spots (represented by the Longevity Hill and *Kunming* Lake). Among them, scenic spots can be subdivided into three areas: the front range of Longevity Hill, *Kunming* Lake, and the back of Longevity Hill and the back of *Kunming* Lake. The front range of Longevity Hill originally was named *Wengshan* (Jar Hill), *Jinshan* (Gold Hill). The south slope which is close to the *Kunming* Lake, together with the beautiful scenery of the lake, makes a beautiful

• 佛香阁

佛香阁是一座宏伟的塔式建筑，为颐和园建筑布局的中心。

Tower of Buddhist Incense (*Foxiang Ge*)

Tower of Buddhist Incense, a magnificent tower, is the center of the overall layout of the Summer Palace.

颐和园全园景观大致分为三个区域：政治活动区（以仁寿殿为代表）、生活区（以乐寿堂、玉澜堂、宜芸馆为代表）、风景游览区（以万寿山、昆明湖为代表）。其中，风景游览区又分为万寿前山、昆明湖、后山后湖三个区域。前山景区的万寿山原名"瓮山""金山"，南坡（即前山）依昆明湖，湖光山色，形成优美的自然风景画卷。佛香阁耸立其上，东西南北各分立着两条垂直对称的建筑轴线：长廊为东西轴线，从长廊中部起为

• 长廊
长廊全长728米，是中国园林中最长的廊，如彩带一般把万寿山前山各景点连接起来。

The Long Corridor
The Long Corridor is 728 meters long, and is the longest corridor in Chinese gardens. It looks like a colorful belt which links together all the scenic spots around the Longevity Hill.

picture of natural scenery. On top of the hill is *Foxiang Ge* (Tower of Buddhist Incense). From the top of *Foxiang Ge*, two vertical and symmetric axes of buildings can be seen. The Long Corridor is the east-west axis and the south-north axis starts from the midpoint of the Long Corridor. Along this south-north axis are *Paiyun* Gate, the Second Palace Gate, *Paiyun* Hall, *Dehui* Hall, *Foxiang Ge* and *Zhihuihai* (Sea of Wisdom Temple). The east-west axis and the south-north axis together form a complete space that is full of changes. *Foxiang Ge* is 36 meters high, which is the biggest building in the garden. It is octagonal with the exterior eave of four layers and the interior eave of three layers. On both sides of the central axis of the central complex of buildings are *Wufang* Pavilion, *Qinghua Xuan*, *Zhuanlunzang* Pavilion (Pavilion of Scriptures Archive), and *Jieshou* Hall. On the west slope are *Huazhongyou*, *Yunsongchao*, and *Tingli Guan* (Hall of Listening to Orioles); on the east slope are *Wujinyi Xuan* (House of Endless Consciousness), *Yangyun Xuan*, *Leshou* Hall, *Yiyun* House, *Yulan* Hall and *Renshou* Hall, etc.

The *Kunming* Lake covers about three quarters of the total area of the Summer Palace. This area is beautiful and gives

南北轴线，分列着排云门、二宫门、排云殿、德辉殿、佛香阁，直至智慧海，是完整又具变化的空间序列。佛香阁为平面八角形，外檐四层、内檐三层，通高36米，是园内最大的建筑物。中央建筑群的中轴线两侧分别为五方阁、清华轩、转轮藏、介寿堂。前山西坡上为画中游、云松巢、听鹂馆。东坡上为无尽意轩和养云轩、乐寿堂、宜芸馆、玉澜堂、仁寿殿等。

tourists a fine and open view, which is the perfect spot for enjoying the beauty of the lake.

The West Dyke is an embankment stretching south-north across *Kunming* Lake, which is a scenic sight. It imitates the *Su* Causeway on Hangzhou West Lake and weeping willows are planted as well. The beautiful scenery of West Dyke echoes that of *Yuquan* Hill outside the Summer Palace and the *Yufeng* Pagoda on the hill, displaying the scenic beauty of green trees

- 柳桥

柳桥在西堤南端，是一座亭桥。柳桥与西堤一同镶嵌在湖水之中，非常壮丽。

The Willow Bridge

Located at the southern end of West Dyke, the Willow Bridge is a pavilion bridge. The bridge and the embankment seem to be embedded into the lake, which is a brilliant scene.

昆明湖的面积约占颐和园总面积的3/4，整个景区视野开阔，环境优美，是观赏湖光山色的绝佳之地。

西堤为南北纵穿于昆明湖之上的一道堤岸，是湖上较为优美的风景，模仿杭州西湖的苏堤，种植垂柳，自然景色优美，与园外的玉泉山和山顶的玉峰塔相映成趣，呈现出一派绿树浓荫、湖光潋滟的江南美景。

后山后河区的景观主要为四大部洲、谐趣园、苏州街、通云城关、妙觉寺、清琴峡等。景区内绿草茵茵、古木葱郁、山环水绕，古雅幽静。

and glittering lake in southern China.

The major scenic spots at the back of Longevity Hill and the back of *Kunming* Lake are the Four Great Lands, *Xiequ* Garden (Garden of Harmonious Interests), *Suzhou* Street, *Tongyunchengguan* (Entrance to the Clouds), *Miaojue* Temple, and *Qingqin* Gorge, etc. In this area, green grass, lush old trees, hills and waters together make a quite elegant view.

- **十七孔桥**
  十七孔桥是一座联拱石桥，共有17个桥洞，长150米，堪称中国园林中最大的桥梁，桥洞与水中的倒影连成纺锤形，十分美丽。

  Seventeen-arch Bridge (*Shiqikong Qiao*)
  Seventeen-arch Bridge is a joint arched bridge with 17 arches. It is 150 meters long, and is known as the longest bridge in Chinese gardens. The arches and its reflections in the water connect with each other and form a very beautiful spindle shape.

避暑山庄

　　避暑山庄原名"热河行宫",是现存最大的行宫御苑,位于河北省承德市北部,是清朝皇帝夏日避暑、处理朝政之地。避暑山庄总占地564万平方米,历时87年建成,相当于颐和园面积的两倍,为中国最大的古典皇家园林。

　　避暑山庄的总体布局与建筑设计利用自然之景,山中有园、园中有山,颇具自然野趣。山庄的建筑布

- 承德避暑山庄示意图
Sketch Map of Chengde Mountain Resort

The Mountain Resort in Chengde

Originally named *Rehe* Palace, the Mountain Resort is the biggest existing imperial resort garden. It is located in the north of Chengde City in Hebei Province. The Mountain Resort is where emperors of the Qing Dynasty spent summer days and handled political affairs. It took 87 years to complete the resort. Covering a total area of 5,640,000 square meters, the Mountain Resort is twice the size of that of the Summer Palace. The Mountain Resort is the largest classical royal garden in China.

　　The Mountain Resort's overall layout and architectural design borrows natural scenery to build a landscape of gardens within mountains and mountains within gardens, possessing a characteristic of natural charm. The architectural layout of the resort can be divided into two regions: the region of palaces and the region of gardens. The region of gardens can be further divided into the lake area (the southeastern part), the mountain area (the northwestern part), and the plain area (the northeastern part). The region of palaces is situated at the southern edge of the resort. It is where emperors lived, read, handled daily administrative affairs, and had entertainment. This region is a

• **青莲岛**

此岛位于如意洲北澄湖内，岛上主体建筑为仿浙江嘉兴烟雨楼而建的烟雨楼，是皇帝赏景的地方。

Cyan Lotus Islet

The Cyan Lotus Islet is located in the northern *Cheng* Lake of *Ruyi* Island. The major building on the islet is the House of Mists and Rains which is a copy of the Tower of Mists and Rains at Jiaxing, Zhejiang Province. The House of Mists and Rains is where emperors appreciated the scenery of mists and rains.

局分为宫殿区、苑景区两部分。苑景区又分为湖区（东南部）、山区（西北部）、平原区（东北部）。宫殿区建于南端，为皇帝处理日常政务、居住、读书、娱乐的场所，包括正宫、松鹤斋、万壑松风、东宫，为平行的建筑群。避暑山庄东南部为湖区，是苑景区的精华，这里集中了半数以上的建筑，运用对比、借景、障景等造园手法，实现景与景之间的连接、组合。

complex of parallel buildings, including the Main Palace, *Songhe Zhai*, the House of *Wanhesongfeng*, and the East Palace. The southeastern part of the Mountain Resort is the lake area, which is the essence of the entire region of gardens. More than half of the buildings are located here. With the application of landscape gardening techniques like comparison, borrowed scenery and obstructive scenery, the connection and combination of different scenic spots are achieved.

The lake area in the southeastern

东南部的湖区面积49.6万平方米,有大小湖泊8处,即西湖、澄湖、如意湖、上湖、下湖、银湖、镜湖及半月湖,统称为"塞湖"。湖区建筑风格多摹自江南风景名胜,如烟雨楼即是模仿浙江嘉兴南湖烟雨楼而建。

part covers an expanse of 496,000 square meters. There are eight lakes of different sizes: the West Lake, the *Cheng* Lake, the *Ruyi* Lake, the Upper Lake, the Lower Lake, the Silver Lake, the Mirror Lake and the Half Moon Lake. These lakes as a whole were called *Sai* Lake. The architectural styles of the lake area are mostly modeled on the famous landscape gardens of southern China. For instance, the House of Mists and Rains is a copy of the Tower of Mists and Rains at *Nanhu* Lake in Jiaxing, Zhejiang Province.

The northern part of the Mountain Resort is a plain area. Few buildings can be found here. This area is mainly dominated by grasslands and forests, giving out a strong flavor of areas beyond the Great Wall. The northwestern part is the mountain area which covers two thirds of the total area of the resort. Ranging from

• **永佑寺舍利塔**

此塔位于万树园东北侧,建于清乾隆十六年(1751年),通高65米,是乾隆帝下旨仿照杭州六和塔而建。

Sarira Stupa in *Yongyou* Temple

The Sarira Stupa is located in the northeastern side of the Garden of Millions of Trees. The 65 meters high Sarira Stupa was built in 1751 (the 16th Reign Year of Emperor Qianlong of the Qing Dynasty) at the decree of Emperor *Qianlong* that it should imitate *Liuhe* Pagoda (Pagoda of the Six Harmonies) in Hangzhou.

避暑山庄北部为平原区，建筑较少，主要为草地、树林，塞外风情浓郁。山庄西北部为山区，占山庄面积的2/3。四条沟壑自南向北依次为榛子峪、松林峪、梨树峪、松云峡。这四条峡谷可登临、游览，其间林木茂盛、古树参天。建筑置于山顶，设计巧妙适于观景，颇具自然野趣。

south to north are 4 successive valleys: *Zhenzi Yu* (Hazels Valley), *Lishu Yu* (Pear Trees Valley), *Songlin Yu* (Pine Trees Valley) and *Songyun Xia* (Pine Clouds Gorge). One can climb and tour in the four valleys where lush forests and towering old trees can be seen everywhere. Buildings are smartly built on top of the mountain, which is suitable for sightseeing, creating a natural sense of rustic charm.

- 冷香亭
 Lengxiang Pavilion (Chilly Fragrance Pavilion)

> 私家园林

中国的私家园林是相对于皇家园林而言的，其所有者和经营者多为历代文人、官宦、富商。最早的私家园林大约出现在西汉时期，到隋唐时期发展成熟，而宋代是私家园林发展的第一个高峰期，受到中国山水画崛起的影响，宋代的私

> Private Gardens

As opposed to royal gardens, Chinese private gardens are mostly owned by literati, officials and wealthy businessmen. Private gardens first appeared in the Western Han Dynasty (206 B.C.-25 A.D.), developed and matured during the Sui and Tang dynasties. The Song Dynasty (960-1279) witnessed the first peak of the development of private gardens. Under the influence of the rise of Chinese landscape

- 东汉画像砖拓片

这件画像砖出土于四川成都，画中表现了一座完整的住宅建筑群，呈两路跨院，左边跨院的庭院中蓄养着供观赏的禽鸟，右跨院后方的一个较大的院落就是宅中花园，园东南角建有一幢高大的阙楼。

Rubbing of Painting on a Brick of the Eastern Han Dynasty (25-220)

This piece of brick of the Eastern Han Dynasty (25-220) was unearthed in Chengdu, Sichuan Province. The painting performs a complete residential building complex which has two courtyards. In the yard of the left courtyard, birds were kept for appreciation; another larger yard at the back of the right courtyard is a garden. At the southeastern corner of the garden there is a tall *Que* building (Watch Tower).

家园林更趋于精雅,出现了很多名园。明清两代,尤其是从清代乾隆时期到清末,私家园林的建造达到了第二个高峰,现存许多著名的私家园林就建造于这一时期。文人雅士以风雅高洁自居,将诗情画意融贯于园林设计之中。

私家园林规模比皇家园林要小,位置多集中于城市中或城市近郊,受地域面积的限制,在布局上特别强调因地制宜,风格精雅别致。就现存的私家园林而言,南方园林无论是在数量上还是在艺术上都胜于北方,其中尤以苏州园林的艺术成就最高。而晚清时期,岭南园林异军突起,与北方、江南的私家园林形成鼎立之势。北方园林、江南园林、岭南园林,代表了中国私家园林的主流。

paintings, private gardens in the Song Dynasty became more exquisite, and many famous gardens came into being. In the Ming and Qing dynasties, especially from the Qianlong Period to the end of the Qing Dynasty, the construction of private gardens reached its second peak. And many existing famous private gardens were built in this period. Considering themselves to be noble and elegant, literati embedded a poetic and artistic sense into private gardens.

Private gardens are smaller than royal gardens. Most private gardens are located in urban and suburban areas. Due to the limitation of construction area, the layout of private gardens specially emphasizes the local conditions, and the style is elegant and exquisite. Among existing private gardens, southern private gardens excel those in northern China, whether in quantity or in landscape gardening. In particular, the gardens in Suzhou scored the highest artistic achievement. During the late Qing Dynasty, *Lingnan* gardens developed quickly, and soon became as famous as the gardens in the Yangtze Delta and northern China. The gardens in these three areas represent the mainstream of Chinese private gardens.

北京恭王府花园

　　北方的私家园林主要分布在北京、天津、河北、山东、山西等地，规模比一般的宅院要大，建筑

- 北京恭王府花园诗画舫

  诗画舫是位于花园西部的一座敞厅，建于湖中，四面环水，是昔日园主人赏荷垂钓的地方。

  Hall of Poetry and Painting in Prince Gong's Mansion, Beijing

  The Hall of Poetry and Painting is an open hall situated in the west of the garden. It is built on the lake and is surrounded by water. The Hall of Poetry and Painting used to be the place where the owner appreciated lotus flowers and went fishing.

Prince Gong's Mansion in Beijing

Private gardens in northern China are mainly located in Beijing, Tianjin, Hebei, Shandong, Shanxi and other places. The scale is larger than that of average houses, the architectural image is strong and stable, and the space is open and broad, which give these private gardens a characteristic of northern China: strong and powerful. Prince gardens are special in northern private gardens. Compared with common private gardens owned by officials and literati, prince gardens contain more buildings, are larger in area, and have rather neat and regular layout. Today only the Prince Gong's Mansion in Beijing is relatively well-preserved.

Prince Gong's Mansion, also known as *Cuijin* Garden (Beauty Gathering Garden), was originally the former residence of He Shen who was an important minister during the Qianlong Period. Later it was forfeited by the Qing government and was changed to a prince mansion. The mansion melts styles and layouts of gardens in southern and northern China into one, and absorbs architectural elements of both China and the West. The buildings of the mansion can be divided into two parts: the mansion house and the garden. The mansion house is in front of the garden. The garden covers 28,000 square meters, and totally there are 31 ancient buildings. The layout of the whole garden is elegant and beautiful, full of subtlety and can be divided into three parts: the middle axis, the eastern axis and the western axis.

A variety of landscape gardening techniques are adopted in the construction of Prince Gong's Mansion to distinguish the garden and the mansion house. Entering the garden gate, one can see rockeries made by bluestones on both sides; flowers and trees are planted on the rocks; paths

恭王府花园妙香亭

妙香亭位于花园东边，为上下两层。上层以八根柱子支撑平顶，呈盛开的海棠花式，亭子周围原来种满丁香花，花开时芳香四溢，所以名为"妙香亭"；下层为方十字形，原名"般若庵"。这种上圆下方的设计代表了中国古代"天圆地方"的宇宙观。

Miaoxiang Pavilion in Prince Gong's Mansion

Miaoxiang Pavillion (Pavilion of Wonderful Fragrance) is situated in the east of the garden. It's a two-storey building. The upper storey is supported by eight columns, in the round shape of a full bloom of flowering Chinese crab-apple. Originally the pavilion is surrounded by lilac trees. When flowers bloomed, fragrance was everywhere. That's why the pavilion is named *Miaoxiang* Pavilion (Pavilion of Wonderful Fragrance). The lower storey is in the shape of a square cross, formerly known as "Wisdom Convent". This kind of design—the upper is round and the below square—represents the ancient Chinese view of the universe that "heaven is round and the earth square".

形象稳重敦实，园林空间开朗，具有雄浑大气的特点。王府花园是北方私家园林的一个特殊类别，与一般官僚、文人的私家园林相比，其建筑数量多、体量大、院落布局规整。现在仅有北京恭王府花园保存较为完整。

恭王府又称"萃锦园"，原为清乾隆年间重臣和珅的私宅，后被朝廷没收改作王府。此园融江南园林风格与北方建筑格局为一体，汇中西建筑元素于一园。建筑分为府邸和花园两部分，府在前而园在后。花园占地2.8万平方米，有古建筑31处，全

wave between rockeries. The two rockeries respectively extend toward the east and the west, and with rolling hills they create a wild imagery of hills and forests. Outside the eastern and the western part of the garden, hills separate the garden from the noisy outside world, forming a relatively closed environment.

• 安善堂
安善堂是一座宽敞大厅，是花园中路最主要的建筑，建筑风格古朴典雅，当时恭亲王常在此设便宴招待客人。

Anshan Hall

The *Anshan* Hall (Hall of Peace and Goodness) is a spacious and open hall. It is the major building on the central axis of the garden, which is simple and elegant in its architectural style. Prince Gong used to entertain his guests with informal dinners here.

园布局分中、东、西三路，景致幽雅秀丽，设计极尽精妙。

恭王府花园造园者采用了多种手法，使花园与王府的住宅区别开来。一进园门，即在两侧用青石垒造假山，山上植以花木，山间留出山径，两山向东西延伸，山峰连绵，具有山林野趣的意境；另外在东西两路的外侧用土山与园外的喧嚷相隔离，形成相对封闭的园林环境。

在中路与东路几重规则的院落中，建造了形状不规则的水池、叠山、块石，用满院竹林和古松、古槐等乔木以及各种灌木和花草，打破了由于布局规整而形成的呆板与严肃。建造者还在西路专门设计了以山水为主体的园林景观，布局疏朗，给人一种清新的感觉。这使得恭王府花园既具有皇族气魄，又不失自然山水情趣。

苏州拙政园

江南私家园林以苏州园林成就最高。拙政园是江南园林的代表，也是苏州园林中面积最大的山水园林。拙政园总面积4.1公顷，分为东、中、西三部分。

In the several courtyards in regular shapes on the central and the eastern axes, irregular shaped pools, stacked mountains and stones are constructed, and bamboos, old pines, old locust trees and other trees, bushes and flowers are planted. This breaks the rigidity and seriousness of the regular layout. Constructors also specially design a landscape that focuses on hills and water on the western axis. The layout is relatively loose, bringing in a sense of freshness. All these enable the Prince Gong's Mansion to own the spirit of the royal family and the pleasure of natural landscape.

Humble Administrator's Garden in Suzhou

Suzhou Private Gardens witness the highest achievement of private gardens in southern China. The Humble Administrator's Garden is the representative of gardens in southern China and the largest landscape garden in Suzhou. Covering a total area of 4.1 hectares, the Humble Administrator's Garden can be divided into three parts: the eastern, the central and the western parts.

The central part is where the major buildings of the garden lie, and the overall layout focuses on ponds. Waterside buildings are distributed in a beautiful

中部是拙政园的建筑主体，总体布局以水池为中心。临水建筑错落有致、主次分明，保持了明清园林质朴、疏朗的风格。远香堂为景区主体建筑，位于水池南岸，与东西两山岛隔池相望，池水清澈广

and orderly way, which keep a simple and spacious style in the Ming and the Qing dynasties. As the major building of the garden, *Yuanxiang* Hall (the Hall of Distant Fragrance) is located on the south embankment of the pond, facing the two islets in the east and the west across the pond. The pond is broad, the water in the pond is clear, and lotus flowers are planted everywhere. *Yuanxiang* Hall and *Xuexiang Yunwei* Pavilion (The Prunus Mume Pavilion) across the water form the oppositve scenery, which together constitute the south to north axis of the central part of the garden.

In the pond two islet are constructed in the east and the west, which separate the pond into the southern and the northern landscape spaces. The western islet is larger and the rectangular *Xuexiang Yunwei* Pavilion is on top of the islet; at the back of the eastern islet is the hexagonal *Daishuang*

- 苏州拙政园松风亭

Songfeng (Pine Wind) Pavilion in the Humble Administrator's Garden in Suzhou

阔，遍植荷花。远香堂与雪香云蔚亭隔水互成对景，共同构成中部园区的南北中轴线。

水池之中分别构筑东、西两个岛山，把水池划分为南北两个景观空间。西山较大，山顶建长方形的雪香云蔚亭；东山山后建六方形的待霜亭；西山的西南角建六方形的荷风四面亭。

Pavilion and at the southeast corner of the western islet is *Hefeng Simian* Pavilion (Pavilion in the Lotus Breezes).

Yiyu Xuan, which is on the west side of *Yuanxiang* Hall, and *Xiangzhou* (the Boat-shaped Building of the Fragrant Islet) to the west of *Yuanxiang* Hall face each other from a distance. *Yiyu Xuan* and *Xiangzhou*, together with *Hefeng Simian* Pavilion in the north stand prominently, from where tourists can enjoy the scenery of lotus flowers from different perspectives. A stream waves from the west of *Yiyu Xuan* into the house in the north where three pavilions stretch across the water, which together are named *Xiao Canglang* Pavilion (Pavilion of Small Surging Waves). *Xiao Canglang* Pavilion and the Small Flying Rainbow Bridge in the north constitute a quiet waterscape courtyard.

In addition, the central part contains *Chengguan* Tower, *Yulan* Hall, *Jianshan*

- 苏州拙政园"柳阴路曲"曲廊
  Liuyinluqu (Shade of Willow Tree and Winding Path) Zigzag Corridor in the Humble Administrator's Garden in Suzhou

- 卅六鸳鸯馆

这是拙政园西花园的主体建筑，采用"鸳鸯厅"的建筑形式，南部为"十八曼陀罗花馆"，北部为"卅六鸳鸯馆"。

The Hall of 36 Pairs of Mandarin Ducks

These are the major buildings in the western part of the Humble Administrator's Garden, which adopt the architectural style of *Yuanyang* Hall (a couple of halls). The southern hall is the Hall of 18 Datura, and the northern hall is the Hall of 36 Pairs of Mandarin Ducks.

- 兰雪堂

兰雪堂是拙政园东部园区的主厅，始建于明崇祯八年（1635年），环境幽僻。

Lanxue (Orchids Snow) Hall

Built in 1635 in the Ming Dynasty (1368-1644), the *Lanxue* Hall is the main hall in the eastern part of the Humble Administrator's Garden. It is secluded and in a tranquil environment.

- 苏州拙政园示意图

Sketch Map of the Humble Administrator's Garden in Suzhou

- 香洲

香洲是拙政园中的标志性景观之一，为典型的"舫"式结构，有两层舱楼，通体高雅而洒脱。

Xiangzhou

Xiangzhou is one of the iconic landscapes in the Humble Administrator's Garden. It is in a typical boat shape with two cabin floors. The entire landscape looks elegant.

- 与谁同坐轩

与谁同坐轩依水而建，平面为扇形，屋面、轩门、窗洞、石桌、石凳及轩顶、灯罩、墙上匾额、半栏均呈扇面形。

The "With Whom Shall I Sit" Pavilion

The "With Whom Shall I Sit" Pavilion is built near the water. The plane of the pavilion is a fan shape. The roof, the door, windows, stone tables, stone stools and the top, the lampshade, plaques on the wall and also the half-rails are all in the shape of a fan.

远香堂西侧的倚玉轩与其西边的香洲遥遥相对，而二者与北面的荷风四面亭又呈三足鼎立之势，游人可变换不同角度欣赏荷景。倚玉轩西侧有一曲水湾深入南部住宅，里面有三间水阁横架水面，名为"小沧浪"，它和北面的廊桥小飞虹共同构成幽静的水景庭院。

此外，中部园区还有澄观楼、玉兰堂、见山楼、枇杷园等，枇杷园中又建玲珑馆和春秋佳日亭。而见山楼三面环水，为园区重要的观景点，经爬山廊直达楼上，可遥望对面的雪香云蔚亭、南轩和香洲景观。

西部园区又称"补园"，主要建筑为卅六鸳鸯馆，是当时园主人宴请宾客的场所。卅六鸳鸯馆周围的水池呈曲尺形，临水建筑扇面形亭"与谁同坐轩"在扇面两侧实墙上开两个扇形空窗，分别对应倒影楼和卅六鸳鸯馆，后窗又正好映入山上的笠亭，与笠亭的顶盖恰好配成一个完整的扇形。其他建筑还有留听阁、宜两亭、倒影楼、水廊等。

Tower (Mountain in View Tower) and the Garden of Loquat, etc. In the Garden of Loquat are the *Linglong Guan* (Exquisite Hall) and the Pavilion of Spring and Autumn. With its three sides surrounded by water, *Jianshan* Tower is the major viewpoint in the garden. Through a corridor tourists can climb up to the building and see the landscape of the distant *Xuexiang Yunwei* Pavilion, the South *Xuan* and *Xiangzhou* in the opposite.

The western part of the garden is also known as the Supplementary Garden. Its major building, the Hall of 36 Pairs of Mandarin Ducks is where the owner of the garden entertained his guests. The pool around the hall is L-shaped. The waterscape building, "With Whom Shall I Sit" Pavilion, is in a fan shape. On both sides of the fan-shaped wall there are two fan-shaped windows, which respectively echo the Reflection Building and the Hall of 36 Pairs of Mandarin Ducks. The window at the back incidentally corresponds to the *Li* Pavilion on the hill, and together with the roof of the *Li* Pavilion a complete fan shape is formed. Other buildings are *Liuting Ge* (Stay and Listen Pavilion), *Yiliang* Pavilion, the Reflection Building, and the Water Corridor, etc.

东部原称"归田园居",为1959年重建,仍保持疏朗明快的风格,主要建筑有兰雪堂、芙蓉榭、天泉亭、缀云峰等。

番禺余荫山房

余荫山房,又名"余荫园",位于广东番禺南村镇东南角,是清道光年间举人邬燕山为纪念其祖父

- **余荫山房浣红跨绿桥**（图片提供：图虫创意）
浣红跨绿桥横卧在莲池之上,被誉为岭南园林最经典的造型,成为余荫山房的标志性景观。

The Bridge of *Huanhong Kualv* in *Yuyin Shanfang*

The Bridge of *Huanhong Kualv* stretches above the lotus pool. Known as the most classical design of *Lingnan* gardens, the bridge is an iconic scenic spot of *Yuyin Shanfang*.

Originally the eastern part was known as *Guiyuantianju* (Dwelling upon Return to the Countryside). Though rebuilt in 1959, it remains the bright and spacious style. Major buildings in this area are *Lanxue* Hall, *Furong* Pavilion, *Tianquan* Pavilion and *Zhuiyun* Mountain, etc.

Yuyin Shanfang in Panyu

Yuyin Shanfang (Mountain Cottage of Abundant Shade), also known as *Yuyin* Garden, is located at the southeast corner of Nan Villiage in Panyu, Guangdong Province. *Yuyin Shanfang* is a private garden built by Wu Yanshan who was a successful candidate in the imperial examinations at the provincial level in the Daoguang Period, to commemorate his grandfather Wu Yuyin. It was built in 1867 (the Sixth Reign Year of Emperor Tongzhi of the Qing Dynasty) and completed in 1871. The garden is famous for its unique style of delicacy, which is one of the four famous gardens in Guangdong.

Yuyin Shanfang absorbed the architectural style of gardens in Suzhou and Hangzhou. The layout of the entire garden is exquisite and elegant. Buildings like

邬余荫而建的私家花园，始建于清同治六年（1867年），于同治十年（1871年）建成。该园以小巧玲珑的独特风格著称于世，为广东四大名园之一。

余荫山房吸收了苏杭庭院建筑艺术风格，整座园林布局灵巧精致，在有限的空间里分别建筑了深柳堂、榄核厅、临池别馆、玲珑水榭、来薰亭、孔雀亭和廊桥等，在面积并不大的山林里，浓缩了园林的主要设施和景致，在有限的空间中注入了幽深广阔的无限佳景。

余荫山房占地面积1598平方米，坐北朝南，以廊桥为界，将园林分为东、西两个部分。西半部以长方形的石砌荷池为中心，池南有造型简洁的临池别馆，池北为主厅深柳堂。深柳堂是园中主体建筑与精华所在，堂前两壁隔扇窗古色古香，厅上两幅花鸟花罩栩栩如生，侧厢三十二幅桃木扇格画橱、碧纱橱的几扇紫檀屏风都是著名的木雕珍品。

东半部的中央为一八角形水池，池中有八角亭一座，名"玲珑水榭"。水榭东南沿着园墙布置了

Shenliu Hall, *Lanhe* Hall, *Linchi Bieguan*, *Ling Long* Water Pavilion, *Laixun* Pavilion, Peacock Pavilion and *Langqiao* (covered bridge) are built in a limited space; major facilities and landscapes are concentrated in a not-large forested mountain. In this way, vast and infinite landscapes are injected into a limited space.

Facing south, *Yuyin Shanfang* covers an area of 1,598 square meters. The *Langqiao* separates the garden into two sections: the eastern section and the western section. A rectangular lotus pool made by stone is the center of the western section. In the south of the pool is the *Linchi Bieguan* which is simple in design, and the north of the pool is the main hall — the *Shenliu* Hall. *Shenliu* Hall is where the major buildings and the essence of the garden lie. In the front of the hall are two antique windows, and in the hall there are two flowered covers with vivid images of birds and flowers. The thirty two fan-shaped painting cabinets made by peach wood, and the several screens made by red sandalwood of *Bisha Chu* (*Bisha* Lattice Doors) are famous and precious wood carving works.

The center of the eastern section is an octagonal pool on which there is an octagonal pavilion named *Linglong* Water

假山，水榭东北点缀着挺秀的孔雀亭和半边亭（来薰亭）。周围还有许多株大树菠萝、蜡梅花树、南洋水杉等珍贵古树。东西两半部的景物通过一座名叫"浣红跨绿"的拱桥有机地结合在一起。

- 余荫山房玲珑水榭内景（图片提供：图虫创意）

  玲珑水榭的平面为八角形，四面环水。室内八面都设有明亮的细密花格长窗，图案在烦琐之间有种谐调的美感。

  the Inner Scene of *Linglong* Water Pavilion in *Yuyin Shanfang*

  The plane of the Water Pavilion is an octagon and is surrounded by water. The eight sides indoors are all decorated with bright long windows with small and flower-shaped lattices. A sense of harmonization is achieved in the complexity of the patterns.

Pavilion. In the southeast of the pavilion rockeries are made along the wall of the garden, and the northeast of the Water Pavilion is dotted with the beautiful Peacock Pavilion and the Half Side Pavilion (*Laixun* Pavilion). Around the Water Pavilion many precious old trees are planted, such as Artocarpus heterophyllus Lam trees, calyx canthus trees and Southern Asia metasequoia trees. With an arch bridge named *Huanhong Kualv* (literally to wash the red and go across the green), the landscape in the eastern and the western sections are organically connected together.

碧纱橱

碧纱橱是清代建筑内屋中的隔断，类似落地长窗，落地长窗通常多安装在建筑外檐，而碧纱橱主要用在内屋，做工更为精细，在南方称"纱隔"。碧纱橱通常用于室内进深方向的柱间，起到分隔空间的作用。碧纱橱经常整排使用，一般由四至十二扇隔扇组成，除中间两扇能开启外，其余均为固定扇。碧纱橱通常饰以浮雕图案和诗文绘画，具有很强的装饰作用。

Bisha Chu

Bisha Chu (*Bisha* Lattice Doors) is the separation in the inner house in the Qing Dynasty (1616-1911). It is similar to the French window. The French window is usually installed in the exterior of the building while *Bisha Chu* is mainly used in the interior house, and the making of *Bisha Chu* is more refined. In the south it is called *Shage* (Veil Separation). *Bisha Chu* is generally used in inner rooms to separate spaces. Often a row of *Bisha Chu* is used together. Generally *Bisha Chu* is composed of four to twelve lattice doors. Except that the two lattice doors in the middle can be opened, the others are fixed doors. *Bisha Chu* is usually adorned with embossments, poems and paintings to achieve a highly decorative effect.

• 苏州网师园看松读画轩内的碧纱橱
Bisha Chu in the *Kansong Duhua Xuan* (Watching Pines and Appreciating Paintings Studio) in the Garden of Master-of-nets in Suzhou

> 自然景观园林

　　自然景观园林是利用天然山水的局部或片段作为建园的基址，配合周围环境构筑一些园林建筑，经人为加工形成的公共游览性的景观园林。

　　自然景观园林从地形、水源到园内的植物多取自天然，所以具有人工园林无法替代的优势。而且这种园林的营建一般不是在同一时期，而是经过历代的传承，其中的名胜古迹、园林建筑风格多样，人文气息比较浓厚。

　　自然景观园林可以大致分为两种：一种位于城市郊外的自然风光区，规模较大，内容广泛，如浙江杭州西湖、江苏扬州瘦西湖、山东济南大明湖等；还有一种位于城市之内或近郊，观赏游览的主体内

> Natural Landscape Gardens

Natural landscape gardens are public landscape gardens made by human efforts by taking a part or a fragment of natural landscape as the base and constructing garden architecture in the light of their surroundings.

　　Mostly the terrain and water source of natural landscape gardens are from nature, which is a irreplaceable advantage that artificial gardens cannot offer. Moreover, the construction of natural landscape gardens cannot be completed in a short period. Instead, it goes through ages and generations, which forms various scenic spots and brings the garden different architectural styles, and thus has a relatively strong cultural atmosphere.

　　Natural landscape gardens can be roughly classified into two types: one is the natural scenery area located in the outskirts

容是名胜古迹，具有较高的文化价值，人文景观丰富，如浙江绍兴兰亭、浙江嘉兴南湖烟雨楼、江西南昌滕王阁等。

- **《曲水流觞图》局部（明）佚名**

  兰亭位于浙江省绍兴市西南14千米处的兰渚山下。东晋永和九年（353年），大书法家王羲之邀集40多位友人、文士在此举行集会，吟诗作序，成为千古佳话。后人出于仰慕，在兰亭修建了许多纪念性建筑，形成了文化气息浓郁的景观园林。

  Part of *Floating Wine Cups along the Winding Creek* (Ming Dynasty, 1368-1644) Anonymous

  Lanting Pavilion (Orchid Pavilion) is at the foot of *Lanzhu* Mountain, fourteen kilometers southwest of Shaoxing City, Zhejiang Province. In 353A.D. (The Ninth Reign Year of *Yonghe* in the Esatern Jin Dynasty), the famous calligrapher Wang Xizhi convened over forty friends and literati to compose poems, which has become a legend. To express their admiration, future generations have built many monumental buildings in *Lanting*, which form a landscape garden with a strong cultural atmosphere.

of a city, which is usually in a larger scale and has various landscapes. The Hangzhou West Lake in Zhejiang Province, the Slim West Lake in Yangzhou, Jiangsu Province, and the Daming Lake in Jinan, Shandong Province are such natural scenery areas. Another type of natural landscape garden is usually located within the city or in the suburbs. The major content of the gardens is places of historic interest which enjoy a high cultural value and possess rich cultural landscape. Such places are *Lanting* Pavilion (Orchid Pavilion) in Shaoxing, Zhejiang Province, the Tower of Mist and Rain at Nanhu Lake in Jiaxing, Zhejiang Province, and *Tengwang* Pavilion in Nanchang, Jiangxi Province.

西湖

西湖，位于浙江省杭州市西部，以秀丽的湖光山色和众多的名胜古迹闻名中外，被誉为"人间天堂"。

西湖周长约15千米，水面面积约5.66平方千米。孤山峙立湖中，小瀛洲、湖心亭、阮公墩三个小岛鼎立湖心，苏堤和白堤两条长堤把湖面分为外湖、里湖、岳湖、西里湖和小南湖五个部分。环湖有南高峰、北高峰、玉皇山等高山，对水

West Lake

The West Lake is located in the south of Hangzhou City, Zhejiang Province. It is well-known to the world for its picturesque landscape and many historical relics, and it is known as Paradise on Earth.

The perimeter of the West Lake is about 15 km, and its surface area is about 5.66 square kilometers. The Solitary Hill stands alone in the lake, and three islets, the *Xiaoyingzhou*, *Huxin* Pavilion (Pavilion of Lake Heart), and *Ruangongdun*, lie in the center of the West Lake. The lake is

• 曲院风荷
Wine-Making Yard and Lotus Pool in Summer

域形成合围之势。环湖山峦叠翠，花木繁茂，峰、岩、洞、壑之间穿插着泉、池、溪、涧，青翠的树木丛中点缀着楼阁、亭榭、宝塔、石窟。湖光山色，风景如画。

在以西湖为中心的49平方千米园林风景区内，分布有"苏堤春晓""平湖秋月""曲院风荷""雷峰夕照"等名胜40余处。其中西湖十景历代流传，最为著名。

- 青花西湖景色图瓶（清 康熙）
Blue-and-White Vase with Scenery of the West Lake Pattern (Qing Dynasty, Kangxi Period, 1662-1722)

divided by two long causeways, the *Su* Causeway and the *Bai* Causeway, into five areas: Outer West Lake, Inner West Lake, *Yue* Lake, West Inner Lake and Little South Lake. The West Lake is surrounded by the South Peak, the North Peak, the *Yuhuang* Mountain and other mountains which are covered with green trees and beautiful flowers. Springs, pools, streams and creeks can be seen between peaks, rocks, caverns and gullies. Lush and green trees are dotted with many buildings such as pavilions, pagodas and grottoes. What a picturesque landscape!

In the garden areas of 49 square kilometers which centered the West Lake, there are over forty famous scenic spots, such as Dawn on the *Su* Causeway in Spring, Moon over the Peaceful Lake in Autumn, Wine-making Yard and Lotus Pool in Summer, and *Leifeng* Pagoda in the Sunset. Among these scenic spots, the most well-known are the Ten Scenes of the West Lake which gose through ages.

The Ten Scenes of the West Lake are:

1. Dawn on the *Su* Causeway in Spring

The *Su* Causeway is on the west side of the West Lake, whose southern and northern ends link the South Mountain Road and the North Mountain Road. The 2.8-km-long *Su* Causeway was built when

西湖十景分别是：

1. 苏堤春晓

苏堤在西湖西侧，南北两端衔接南山路与北山路，全长2.8千米，是北宋文学家苏东坡在杭州为官时组织民工开浚西湖，挖泥堆筑而成的。堤上还安排映波、锁澜、望山、压堤、东浦、跨虹6座石拱桥，起伏相间。堤上两边夹种桃树、柳树，风光旖旎。苏堤景色四时不同，晨昏各异，晴、阴、雨、雪均

the great literatus Su Shi in the Northern Song Dynasty came to Hangzhou as an official and organized workers to dredge the West Lake and piled up the mud into the causeway. On the causeway there are six undulating stone arch bridges: *Yingbo* Bridge, *Suolan* Bridge, *Wangshan* Bridge, *Yadi* Bridge, *Dongpu* Bridge, and *Kuahong* Bridge. On both sides of the causeway, peach trees and willow trees are planted, forming a charming sight. The *Su* Causeway has different sceneries in the four seasons, at dawn and dusk. Sunny days, cloudy days, rainy days and snowy days all have their attraction. In particular at dawn in spring, light mist covers the lake like thin and soft cloth; and on the causeway weeping willow trees look like clouds. That's why the causeway is known as Dawn on the *Su* Causeway in Spring.

2. Orioles Singing in the Willows

Orioles Singing in the Willows is located in the southeast bank of the West Lake, close to the *Qingbo* Gate on South Mountain Road. It was originally named *Jujing* Garden, an imperial garden of the Southern Song Dynasty (1127-1279) emperors. There used to be a *Liulang* Bridge, and willow trees were planted along the lake. In spring, dense willow trees wave in the breeze, and orioles sing in

- 三潭印月的小石塔
Small Stone Pagoda of Three Ponds Mirroring the Moon

• 从西湖白堤远眺保俶塔
Baochu Pogoda Overlooked from the *Bai* Causeway of the West Lake

有情趣。尤其是春天早晨，湖面薄雾似纱，堤上烟柳如云，故有苏堤春晓之称。

2. 柳浪闻莺

柳浪闻莺位于西湖东南岸，南山路清波门附近。这里原为南宋皇帝的御花园——聚景园，园中原有柳浪桥，沿湖遍植垂柳，繁密的柳丝在春风吹拂下仿佛碧浪翻飞，浓荫深处时时传来呖呖莺声，因而名为柳浪闻莺。

3. 曲院风荷

曲院风荷原本在苏堤北端的跨

the dense willow trees which are waving in the spring breeze. That's why the garden is named Orioles Singing in the Willows.

3. Wine-making Yard and Lotus Pool in Summer

Originally, the Wine-making Yard and Lotus Pool was below the *Kuahong* Bridge on the northern end of the *Su* Causeway. There used to be a wine-making yard which was a brewing farm of official wine. Lotus flowers were planted everywhere in the garden and the air is filled with the fragrance of the flowers. That's why it is called Wine-making Yard and Lotus Pool in Summer. Today's Wine-making Yard and Lotus Pool is several hundred of times larger than the original one and become an exquisite small-size garden. A vast expanse of water is planted with lotus flowers, and corridors and pavilions for viewing lotus flowers are built near the water, which is in a simple and elegant style.

4. Moon over the Peaceful Lake in Autumn

Moon over the Peaceful Lake in Autumn is at the western end of the *Bai* Causeway. Its three sides are facing water and at the back of the fourth side stands the Solitary Hill. There used to be a Lakeview Pavilion in the Tang Dynasty (618-907). In 1699 (the thirty-eighth year of the Reign

虹桥下，宋代时那里有一家酿造官酒的曲院，里面种满荷花，清香四溢，因此有曲院风荷之说。现在的曲院风荷比原来扩大了数百倍，成为一处布局精巧的小型园林，水上种有大片荷花，傍水建造的赏荷廊、轩、亭、阁，古朴典雅。

4. 平湖秋月

平湖秋月位于白堤西端，三面临水，背倚孤山。唐代，这里建有望湖亭。清康熙三十八年（1699年）改建为御书楼，并在楼前挑出水面铺筑平台，题名为平湖秋月。在平台上眺望西湖，景色非常美丽，尤其是皓月当空的秋夜，更是充满了诗情画意。

5. 三潭印月

三潭印月在西湖三岛之一的小瀛洲周围。岛上有曲桥和造型别致的亭、榭，景观富有层次感，意境深邃。小瀛洲岛南的水面上有3座造型美丽的小石塔。每逢秋夜，皓月当空，在塔内点上灯烛，洞口蒙上薄纸，灯光从中透出，宛如一个个小月亮倒映水中，三潭印月则由此得名。

6. 雷峰夕照

西湖南岸的夕照山上，原有一

of Emperor Kangxi of the Qing Dynasty) it was changed into the Imperial Library Building. A platform was built over the selected water in front of the building and was named Moon over the Peaceful Lake in Autumn. Standing on the platform and having a look at the West Lake, one can see the amazingly beautiful scenery of the lake; especially in a full-moon autumn night, the scenery is full of poetic sense and artistic imagination.

5. Three Ponds Mirroring the Moon

Three Ponds Mirroring the Moon is near the Lesser *Yingzhou*, one of the three islets of the West Lake. On the islet, there are curved bridges and pavilions that is of unique shapes on terrace, forming a landscape with rich imagination. Three small stone pagodas of exquisite shapes stand on the lake in the north of Lesser *Yingzhou*. At every bright night in autumn, candles are lit in the pagodas and the holes are covered with thin paper, then when the light comes out from the holes, it seems that small moons are reflected in the water, hence the site is named Three Ponds Mirroring the Moon.

6. *Leifeng* Pagoda in the Sunset

On the Sunset Hill north of the West Lake there used to be a *Leifeng* Pagoda. When sunset came, the pagoda cast a

座雷峰塔。夕阳西照时，塔影横空，金碧辉煌，雷峰夕照由此而名。雷峰塔与保俶塔隔湖相对，湖上双塔，水中双影，与湖中三岛、苏白二堤相辉映，曾给游人增添无限美感。1924年，雷峰塔倒塌，现在的雷峰塔为近年新建。

7. 南屏晚钟

南屏晚钟是指南屏山下净慈寺的钟声。净慈寺始建于公元954年，

beautiful shadow which was splendid. Hence it was named *Leifeng* Pagoda in the Sunset. On the opposite side of *Leifeng* Pagoda is the *Baochu* Pagoda. The two pagodas stand by the lake and cast two shadows in the water, which echoes the three islets and the *Su* Causeway and *Bai* Causeway, bringing infinite beauty to tourists. Unfortunately, *Leifeng* Pagoda collapsed in 1924 but it was rebuilt in recent years.

• 西湖小瀛洲
Lesser *Yingzhou* of the West Lake

西湖金沙堤上的玉带桥
Yudai Bridge on the Jinsha Causeway of the West Lake

是西湖四大丛林寺院之一。寺前原有一口大钟，每到傍晚，钟声响起，晚景十分迷人。

8. 断桥残雪

断桥是白堤的东起点，位于外湖和北里湖的分水点上，因孤山之路到此而断，故名断桥。中国民间传说《白蛇传》的故事，就发生在这里。旧时石拱桥上有台阶，桥中央有小亭，冬日雪霁，桥上向阳面冰雪消融，阴面却留有残雪，桥面

7. Evening Bell Ringing at the *Nanping* Hill

Evening Bell Ringing at the *Nanping* Hill refers to the sound of the bell in *Jingci* Temple at the *Nanping* Hill. First built in 954 A.D., *Jingci* Temple is one of the four great Buddhist temples around the West Lake. Originally there was a big bell in front of the temple. In the evening, when the bell rang, the scene was very attractive.

8. Remnant Snow on the Broken Bridge in Winter

The Broken Bridge is the eastern starting point of the *Bai* Causeway, right on the watershed point of the Outer Lake and North Inner Lake. Because the road to the Solitary Hill ends here, the bridge is named the Broken Bridge. The Chinese folklore *the Legend of the White Snake* happens right here. In the past there were steps on the stone arch bridge, and at the center of the bridge there was a small pavilion. In winter, when snow stops, the snow on the sunny part of the bridge melts while remnant snow can be seen on northern part of the bridge, and thus the bridge appears to be broken. This unique scenery brings the site the name Remnant Snow on the Broken Bridge.

湖心亭
Huxin Pavilion (Pavilion of Lake Heart)

看上去似断非断，景观奇特，因有断桥残雪之名。

9. 双峰插云

双峰插云位于灵隐路上的洪春桥边。"双峰"指的是天竺山环湖山脉中的南高峰、北高峰。两峰遥相对峙，山雨欲来时，双峰的峰尖忽隐忽现插入云端，景致十分壮观。

10. 花港观鱼

花港观鱼位于苏堤南端。古代，有小溪自花家山流经此处入西湖，所以称花港。宋时，花家山下建有卢园，园内栽花养鱼，风光如画，被画家标上花港观鱼之名。这里原来仅有一碑、一亭、一池和三

9. Twin Peaks Piercing the Clouds

Twin Peaks Piercing the Clouds is located by the side of *Hongchun* Bridge on *Lingyin* Road. The Twin Peaks are the South Peak and the North Peak of the mountains of *Tianzhu* Mountian which are around the West Lake. The two peaks face each other in the distance. When rain is coming, the tops of the two peaks pierce into the clouds now and then, and the scenery is spectacular.

10. Fish Viewing at the Flower Pond

Fish Viewing at the Flower Pond is located at the northern end of the *Su* Causeway. In ancient times, there was a stream flowing through *Huajia* Mountain (*Hua* literally means flower) into the West

亩地，现已建成占地20多公顷的大型公园。游人围拢鱼池投饵，群鱼翻腾水面，追逐争食，红光波音，有色有声，呈现一番鱼乐人也乐的景象。

宏村

在中国的乡村地区，一些坐落在山下河岸的优美村落，也可以看作自然景观园林的一种类型。它们以自然山水为骨架，以散落于村落

Lake, hence it was named the Flowered Pond. In the Song Dynasty (960-1279), at the foot of *Huajia* Moutain there was a *Lu* Garden where flowers were planted and fish were kept. Because of its picturesque scenery, the garden was labeled as Fish Viewing at the Flowered Pond by artists. Although originally there was only one monument, one pavilion, one pond and 0.2 hectare of land, today it has been built into a large park covering more than 20 hectares of land. Tourists gather around the fish pond and feed them while groups

- **安徽宏村月沼**

月沼位于宏村的中心位置，被称为"牛胃"。池水常年碧绿，水平如镜。塘沼四周青石铺展，粉墙青瓦整齐有序地分列池边，蓝天白云倒映水中，形成一幅美丽的皖南风景画。

Yuezhao (Moon Pool) in Hongcun Village, Anhui

Located at the center of the Hongcun village, *Yuezhao* is called the stomach of the bull. Green in all seasons, the tranquil pool water is like a mirror. Around the pool, blue stones were paved, and houses with pink walls and blue tiles were orderly arranged. Together with the reflections of the blue sky and white clouds, they form a beautiful landscape of southern Anhui.

中的民宅房舍、小桥、牌坊等为园林建筑，风光秀丽，散发着古朴、自然、纯净的气息。

- 承志堂

承志堂是宏村内一幢保存完整的大型民居建筑，建于清咸丰五年（1855年），是清末大盐商汪定贵的住宅。整栋建筑为木结构，建筑面积3000余平方米，内部砖雕、石雕与木雕装饰数量繁多，工艺精细，富丽堂皇，有"民间故宫"之誉。

Chengzhi Hall

Chengzhi Hall is a well-preserved, large-size residential building in Hongcun Village. It was first built in 1855 (the fifth Reign Year of Emperor Xianfeng) as the private residence of the famous salt merchant Wang Dinggui of the Qing Dynasty(1616-1911). The wood-structure building covers a total area of over 3,000 square meters. Its many internal brick carvings, stone carvings and wood carvings of the building are exquisite and magnificent. Hence the *Chengzhi* Hall gains the reputation as the "Folk Palace Museum".

of fish jump over the water to chase for food. Different colors of the jumping fish and their sound show a happy scene of the enjoyment of both the fish and the tourists.

Hongcun Village

In the rural areas of China some beautiful villages at the foot of a mountain or by the riverside can be considered to be one type of natural landscape gardens. Taking mountains and water of nature as the basic frame, and the scattered houses, small bridges and memorial gateways and so on as garden buildings, these villages are

安徽徽州地区的很多村落都具备了这样的特征，其中地处黄山西南麓黟县境内的宏村颇为著名。宏村位于黟县桃花源盆地的北缘，四周群峰环绕，大小各异的湖池散布其间。民居建筑古朴简约，古楼、古桥、古亭等公共建筑清古悠远，营造出古色古香的村落氛围。

整个村子呈"牛"形结构布局，巍峨的雷岗为"牛首"，参天古木是"牛角"，由东而西错落有致的民居群宛如庞大的"牛躯"。人们将村子西北的一处小溪开凿成水渠，绕屋过户，九曲十弯，聚村中天然泉水，蓄成一口新月形的池塘，形如"牛肠"和"牛胃"；水渠最后注入村南的湖泊，俗称"牛肚"；绕村溪河上还建有四座桥梁，作为"牛腿"。这种别出心裁的村落水系设计，不仅为村民解决了消防用水问题，而且调节了气温，为居民生产、生活用水提供了方便，创造了一种"家家门前有清泉"的良好环境。

全村现保存完好的明清古民居有140余幢，古朴典雅，意趣横生。"承志堂"富丽堂皇，精雕细刻，

beautiful landscapes, exuding a simple, natural and pure flavor.

Many villages in Huizhou area in Anhui Province bear the same features, among which the Hongcun Village in Qianxian County which is southwest of the Yellow Mountain is a famous one. Surrounded by the peaks around, the Hongcun Village is located at the northern edge of *Taohuayuan* Basin in the Qianxian County with lakes and ponds of different sizes scattering within it. Residential houses here are of primitive simplicity and the old buildings, bridges, pavilions and other public buildings are of a long history, which together create an antique flavor.

The whole village is bull-shaped in layout. The high *Leigang* Mountain is the shape of a bull head, and the tall old trees are the bull horn, and the many well-arranged residential houses scatter from the east to the west are the huge body of the bull. A stream at the northwest of the village is built into a ditch waving between the houses, and together with the natural spring in the village forms a crescent-shaped pond, which respectively symbolizes the intestine and the stomach of the bull. The lake in the south of the village where the ditch ends is known to people as the bull belly. On the stream there are four

可谓皖南古民居之最;南湖书院的亭台楼阁与湖光山色交相辉映,深具传统徽派建筑风格;敬修堂、东贤堂、三立堂、叙仁堂,或气度恢宏,或朴实端庄;再加上村中的参天古木、民居墙头的青藤老树、庭中的百年牡丹……真可谓步步入景,处处堪画,同时也反映了悠久历史所留下的广博深邃的文化底蕴。

bridges which are the bull legs. This kind of special design of waters of the village not only provides the water for the purpose of fire control, but also adjusts the temperature here, and provides convenience for the water of residents' living and producing activities, creating a good environment that there's a clear spring in front of every house.

There are more than 140 well-reserved residences of the Ming and Qing dynasties in the village which are simple, elegant and charming. The *Chengzhi* Hall, magnificent with exquisite carvings, is the top hall of the ancient residences in southern Anhui. Pavilions and buildings of South Lake School and the lake scenery add radiance and beauty to each other, which are all in traditional *Huizhou* architectural style. *Jingxiu* Hall, *Dongxian* Hall, *Sanli* Hall and *Xuren* Hall are magnificent or simple or elegant. These halls, together with the towering old trees, old ivy trees at top of the wall of residences, and the century-old peony in the court, form a landscape where beauty is everywhere and each scene is worthy to be painted, and in the meantime reflects the profound cultural heritage left by the long history.

中国园林的造园艺术
Gardening Art of the Chinese Gardens

  中国园林的营造追求模仿自然，也就是用人工的力量来构建自然的景色，达到"虽由人作，宛自天开"的艺术境界。所以，园林中除了大量的建筑物外，还要配置山石，开凿池塘，栽花种树，同时参照山水画笔意和诗文的情调，构成如诗如画的园景。因此，中国园林是建筑、山池、园艺、绘画、雕刻等多种艺术的综合体。

Construction of the Chinese gardens is modeled on nature. It tends to use artificial power to reproduce natural sceneries, so as to achieve the principle that "The works of men should match that of Heaven". As a result, aside from buildings, in a Chinese garden there also exist rockeries, pools, flowers and trees, which are laid out in a poetic and picturesque pattern. In a word, the Chinese garden is a synthesis of all kinds of arts like architecture, design of rockeries and water sceneries, gardening, painting, sculpture, etc.

> ## 空间布局

中国园林的布局，一般分为规则式、自然式和混合式三大类。

规则式布局又称"几何式""对称式""整形式"，体现的是雄伟、庄严、整齐与对称，强调几何图案美。整个园林平面布局如建筑、广场、道路、水面、花草

> ## Spatial Layout

The spatial layout of the Chinese gardens can be classified into 3 types, namely, the regular type, the natural type, and the mixed type.

The regular type can also be called the geometrical type, the symmetrical type, or the orderly type. It represents magnificence, solemnity, order and

- **北京北海示意图**

北京北海的整体布局继承了古代皇家园林传统布局形式，水面占全园面积一半以上，视野开阔，岛屿置于水中，桥堤连接二岛（琼华岛、团城）。二岛呈中轴线分布，岛上以及岸边分别布置寺院、皇家殿堂等建筑物。全园的主景观为最大岛屿琼华岛上的白塔，以其为中心，建筑物分布四周。

Sketch Map of *Beihai* Park in Beijing

The overall layout of *Beihai* Park has inherited the traditional layout of ancient Chinese royal gardens. Lake area accounts for more than half of the whole park area, which allows for a spacious view for people. There are two islets, *Qionghua* Islet (Jade Islet) and *Tuancheng* (the Rround City), in the lake, and they are linked to the banks by bridges. Buildings on the two islets, such as Buddhist temples and imperial temples, are symmetrically distributed. The main attraction in the park is *Bai Ta* (White Dagoba) on the *Qionghua* Islet, around which, buildings are laid out.

- **苏州网师园示意图**

  网师园是苏州园林中前宅后院布局的典型代表。全园以中央水池为中心，建筑大都临池而建，山石堆叠，树木林立，景致多变，整体布局精妙、紧凑而有层次。

  Sketch Map of The Master-of-nets Garden

  The Master-of-nets Garden is a typical representation of the Chinese gardens, which usually contains two parts: the front living area and the back garden area. The garden takes the central pond as its center, and most buildings are constructed near the pond. With stacked rockeries, verdant trees and varied scenes, the garden appears exquisite, compact and well-arranged.

树木等多按照明显的轴线进行几何对称式布置，追求一种均衡、和谐的美。庄严肃穆的皇家园林、纪念性园林采用规则式布局的较多，如北京北海、天坛等。

自然式布局的园林以模仿再现自然为主，不追求对称的平面布局，立体造型及园林要素布置都比

symmetry, putting emphasis on geometrical beauty. Buildings, squares, roads, water sceneries, plants, to name just a few, are laid out in a symmetrical pattern, in order to demonstrate a balanced, harmonious beauty. Royal gardens and memorial gardens, such as *Beihai* Park in Beijing and the Temple of Heaven, tend to use this type.

The natural type aims to imitate and reproduce nature. It pays little attention to symmetry. Instead, buildings and other elements are distributed irregularly, making it hard to locate each other. The natural type is marked by subtlety, peacefulness, delicacy, and is applicable in places that have mountains, pools and rivers, and undulating terrains. Typical gardens built

较自由，相互关系较为含蓄。这种布局形式较适合有山有水有地形起伏的环境，以含蓄、幽雅、意境深远见长。皇家园林中的颐和园，私家园林中的苏州拙政园、网师园等都是自然式布局的代表作品。

所谓混合式园林，主要指规则式和自然式的布局交错组合，这样的园林一般并没有控制全园的中轴线，只在局部景区、建筑中以中轴对称布局。多结合地形，在地形平坦处，根据规划的需要安排规则式的布局；而在地形条件较复杂的地方，如起伏不平的丘陵、山谷、洼地等，则结合地形规划成自然式布局。

in this type include the Summer Palace in Beijing, the Humble Administrator's Garden and the Master-of-Nets Garden in Suzhou.

The mixed type is a combination of the regular and the natural type. Gardens of this type are not symmetrical on the whole, but symmetrical in some particular area or in some particular buildings. It usually depends on the terrains of a garden. In flat areas, the regular type is usually introduced; while in undulating areas such as in hills, valleys and hollows, the natural type is often adopted.

• 从山顶俯瞰避暑山庄

避暑山庄布局可分为宫殿区和苑景区两大部分，其中宫殿区采取规则式布局，风格严谨规整，而苑景区布局充分利用天然山水地势，充满自然野趣，宫殿与天然景观和谐地融为一体。

A Bird's-eye View of Chengde Mountain Resort

Chengde Mountain Resort can be divided into the palace area and the scenic area. The layout of the palace area is of the regular type, which makes it solemn and orderly. In comparison, the scenic area is in the natural type, which makes full use of natural terrains and is full of natural rustic charms. In this way, palaces and natural sceneries are fused harmoniously with each other.

> 造景手法

中国园林在营造中往往通过对比、衬托、尺度、虚实等一系列艺术手法，运用借景、添景、对景、框景、夹景、漏景、障景等基本造景手段，形成高低错落、疏朗有致、充满节奏和韵味的园林景观空间。

借景

"借"包含借用、依据、凭借等义，园林中的借景首先对自身的

> Gardening Techniques

Techniques such as contrasting, backgrounding, sizing and firm-empty are often used in the Chinese gardens. Principles such as borrowed scenery, added scenery, oppositive scenery, enframed scenery, vista line, leaking through scenery and blocking scenery are frequently introduced to make a well-spaced garden full of rhythm and charm.

Borrowed Scenery

The word "borrowed" here is synonymous with "based on" or "depending on". Borrowed scenery is the principle of "incorporating background landscape into the composition of a garden", whose goal is to expand space and enhance changes.

- 湖南岳阳楼近借洞庭湖（图片提供：图虫创意）
  Jinjie (Near Borrowing) of *Dongting* Lake by *Yueyang* Tower in Hunan

051

中国园林的造园艺术
Gardening Art of the Chinese Gardens

景观加以利用，其次将外部景观借至景点中来，扩大空间、增强变化。

借景可分为近借、远借、邻借、互借、仰借、俯借和应时借等形式。近借，指在园中欣赏园外近处的景物。远借一般用于非封闭式的园林，指在园林中看远处的景物，例如在水边眺望开阔的水面和远处的岛屿。邻借，指在园中欣赏相邻园林或院落内的景物。互借，指两座园林或两个景点之间彼此借

There are several categories in borrowed scenery, such as *jinjie* (near borrowing), *yuanjie* (distant borrowing), *linjie* (adjacent borrowing), *hujie* (mutual borrowing), *yangjie* (upward borrowing), *fujie* (downward borrowing) and *yingshijie* (seasonal borrowing). *Jinjie* means appreciating outside scenery which is closed to the garden. *Yuanjie* means appreciating far-away scenery in an open garden, e.g., looking at a broad pond or a distant islet from the bank. *Linjie* means appreciating scenes in neighboring gardens. *Hujie* means mutual appreciation between two gardens or two sceneries. *Yangjie* means

- **颐和园远借玉泉山**
  颐和园以园外西山群峰和玉泉山上的宝塔为借景，形成山外有山、景外有景的丰富景观。
  Yuanjie (Distant Borrowing) of *Yuquan* Mountain by the Summer Palace
  The Summer Palace borrows the distant *Xishan* Mountain and *Yuquan* Mountain as part of its scene.

资对方的景物。仰借，指在园中仰视园外的峰峦、峭壁或高塔。俯借指在园中的高处俯瞰园外的景物。而应时借，指的是借用某一季节或一天中某一时刻的景物，主要是借天文景观、气象景观、植物四季变化景观和即时的动态景观等。

looking upward at mountains, cliffs, or towers from the garden, while *fujie* means looking downward at views outside the garden. *Yingshijie* refers to borrow a scene in a particular season or at a particular time of a day, usually an astronomical phenomenon, a meteorological phenomenon, a seasonal landscape and other dynamic landscapes.

● 苏州拙政园宜两亭

拙政园西部原为清末张氏补园，与拙政园中部分别为两座园林，西部假山上设宜两亭，邻借拙政园中部之景，一亭尽收两家春色。

The Both Families Pavilion in the Humble Administrator's Garden in Suzhou

The western part of the Humble Administrator's Garden, primarily the Supplementary Garden of the Zhangs' Family, and the central part of the garden are actually two smaller gardens. On a rockery in the western garden is a pavilion called the Both Families Pavilion. One can stand in the pavilion and appreciate sceneries in both gardens.

- 仰借
  Yangjie (Upward Borrowing)

- 苏州拙政园雪景（应时而借）
  Snow Scene in the Humble Administrator's Garden in Suzhou (*Yingshijie*, Seasonal Borrowing)

对景

　　对景，简单地讲就是使园林中的两个景观相互观望，以丰富园林景色，一般选择园内透视画面最精彩的位置，作为供游人逗留的场所，例如亭、榭等。这些建筑在朝向上与远景相向对应，能相互观望、相互烘托。

Oppositive Scenery

Oppositive scenery refers to mutual appreciation between two scenic spots. Under most conditions, the best places for sight-seeing are in some buildings such as pavilions and pavilions on terrace, which work with far-away scenic spots or with each other as oppositive sceneries.

- 苏州退思园闹红一舸舫与水香榭互成对景
  Naohongyige Boat House and *Shuixiang* Water Pavilion in *Tuisi* Garden, Suzhou（Oppositive Scenery）

添景

　　添景是使景观画面更加丰富、完整的艺术手法。如中国园林中的拱桥、平桥、廊桥、曲桥等均可增添园中景色，在视觉上起到丰富空间的作用。园林中窗前檐下的枝干、树叶、花果也可以成为添景要素。

Added Scenery

Added scenery can visually enrich and consummate a garden. Decorative bridges like arch bridge, level bridge, gallery bridge and zigzag bridge can visually enlarge the space. Besides, branches, leaves, flowers and fruits of a tree in front of the window are also important added sceneries.

● 添景
Added Scenery

框景

框景是利用门框、窗框、树框、山洞等将景框在"镜框"中，有选择地摄取优美景色，构成如诗如画的园林景观。中国园林典型的框景实例有苏州网师园殿春簃长窗、苏州留园花窗、颐和园一步一景的长廊等。

Enframed Scenery

Doors, windows, trees and caves can be used as "frames" through which certain picturesque sceneries are enframed. Typical enframed sceneries include *Dianchunyi* Long Window in the Master-of-nets Garden, Suzhou, the caved windows in the Lingering Garden, Suzhou, and the long gallery in the Summer Palace.

• 框景
Enframed Scenery

夹景

　　夹景是在轴线、透视线两侧布置花木、山石、建筑等要素强化和突出主要景物的造景手法。夹景强调主景与两侧景物的关系，起到引导视线和增加园林景观景深的作用，多用于园林河流及道路设计中。

- 西湖云栖竹径
  Bamboo-lined Path at *Yunqi* along the West Lake

Vista Line

Vista Line is a gardening technique, that is, using flowers, trees, rocks, buildings and other elements on both sides of the axis line or the perspective line to highlight the main features. Vista line puts emphasis on the relationship between main features and their surrounding features. It works to navigate the eyesight and lengthen the depth-of-focus, and is applied in the designing of creeks and roads.

漏景

漏景是通过院墙或廊壁上的花窗，将室内外或园内外景致组合在一起的造景手法。漏窗上多雕刻有民族特色的几何图形、葡萄、石榴、修竹等植物以及鹿、鹤等珍禽异兽。

Leaking through Scenery

It is a gardening technique in which caved windows on a garden wall or a gallery wall is used to link sceneries in and outside a room or a garden together. The caved windows are often decorated with geometric patterns, plant designs like grapes, pomegranates, bamboo, and animal designs like deer and cranes.

- **沧浪亭园墙上的花窗**
  苏州沧浪亭有108种变化多端的花窗样式，是苏州园林花窗的典型，也是中国园林运用漏景的典范。
  Caved Windows on the Wall of *Canglang* Pavilion, Suzhou
  There are over 100 kinds of caved windows on the wall of *Canglang* Pavilion. They are not only the most typical caved windows in Suzhou gardens but also the most typical application of leaking through scenery in the Chinese gardens.

障景

　　障景，就是在园林中设置像屏障一样的景物或景观。中国园林设计最忌一览无余，而追求曲径通幽的效果，将风景逐步展开，引领游人细细品味，渐入佳境。而障景就是能制造出这种曲折效果的造景手法，多在园林入口处和空间序列的转折引导处设置山石、植物、景墙等造园要素来处理。障景还可障蔽园中不宜显露的空间景物及不良景观。

Blocking Scenery

Blocking scenery is to set up barrier-like scenic spots in a garden. The Chinese garden designers depreciate taking in everything in a glance the most. Gradual unfolding of views or a winding path leading to a secluded quiet place are preferred. Blocking scenery is therefore used to achieve such effects. Blocking scenery is often seen in entrances and turnings, where rockeries, plants or view walls are applied. It can also help keeping privacy and block unattractive spots.

- **拙政园缀云峰**

  缀云峰立于拙政园东部的兰雪堂前，像一个巨大的屏风，挡住游人的视线。障景是苏州古典园林造园最常用的艺术手法之一。

  Zhuiyunfeng (A Rockery) in the Humble Administrator's Garden

  Zhuiyunfeng is situated in front of *Lanxue* Parlor which is in the east of the Humble Administrator's Garden. This rockery, like a big screen, purposefully blocks off the tourists' sight. Blocking scenery is one of the most frequently used techniques in Suzhou gardens.

> 山石的布置

掇山置石是中国园林独有的一种造园手段。掇山又叫"叠山",指在园林中人工堆造假山为景;置石是在园林中有意识地安置零星山石为景。中国园林的掇山置石方法多借鉴传统的绘画理论。

山石的选择

园林叠山前要选择用石。首先根据园林中造景的要求以及功能的需要来决定用石的种类、形态、大小、色泽、质地,同时还应考虑实际条件,以就近、就地取材为佳。我国园林叠山用石以湖石、黄石类为多,有太湖石、房山石、宣石、黄石、青石等及各种石笋。

> The Arrangement of Rocks

Rock piling and rock layout are specially used in the Chinese gardens, with the former referring to piling stones into a rockery, and the later referring to placing fragmentary stones in accordance with some purpose. Ideas from tradtional Chinese painting are usually borrowed in rock piling and rock layout.

The Selection of Rocks

Rocks should be carefully selected before being piled. Considering practical requirements and functions, rocks are chosen according to their types, shapes, sizes, colors or textures. Meanwhile, obtaining local materials or materials in neighboring counties are recommended. Generally, stones used for piling are mainly Lakeside Stones and Yellow Stones like *Taihu* Stone, *Fangshan* Stone, *Xuan* Stone, Yellow Stone, Bluestone as well as all types of stalagmites.

昆石
Kun Stone

- **昆石**

  昆石产于江苏省昆山市昆山玉峰，因洁白晶莹，玲珑剔透，峰峦嵌空，千姿百态，故又名"巧石""玲珑石"，外地人又称"昆山白石"，与太湖石、雨花石等并称于世，为江南名石之一。

  Kun Stone

  Kun Stone is produced in Kunshan Jade Mountain, Kunshan, Jiangsu province. It is also called *Qiao* Stone (*Qiao* in Chinese means delicate) and *Linglong* Stone (*Linglong* in Chinese means exquistite) because of its crystally white color and exquisite texture. People outside the city of Kunshan also call it Kunshan White Stone, which, together with *Taihu* Stone and *Yuhua* Stone, is recognized as prestigious stone in regions south of the Yangtze River.

太湖石
Taihu Stone

- **留园冠云峰**

  太湖石又称"湖石"，是古典园林叠山常用的石料之一，以太湖一带出产者最为著名，故称"太湖石"。冠云峰高6.5米，是江南园林中最高大的一块湖石。此峰玲珑剔透，集瘦、皱、漏、透于一身，堪称太湖石中的极品。

  Guanyun Rockery in Lingering Garden

  Taihu Stone, also named as Lake Stone, is one of the most frequently used rockery stones in the Chinese gardens. *Guanyun* Rockery, which is as tall as 6.5 m, is the highest Lake Stone in gardens on the Yangtze Delta. The rockery is exquisitely translucent. Besides, it fulfills the four features cherished by the Chinese garden makers, namely, *shou* (being slim), *zhou* (having wrinkled textures), *lou* (having small holes), *tou* (having large and permeable holes), all of which make it one of the best-quality Lake Stones.

瘦、皱、漏、透
The Four Features: *Shou, Zhou, Lou, Tou*

中国古代文人欣赏假山石的标准可以总结为"瘦、皱、漏、透"四个字。"瘦"是指石体修长,这体现了中国古代尤其是宋代以来以清瘦为美的审美时尚。"皱"指的是表面布满凹凸变化的褶皱,这是由千百年地壳运动和风化作用形成的,给人以厚重的沧桑感。而"漏"和"透"是指石体中出现前后、上下贯穿的洞穴,往往大小不等、方向各异,产生各种千变万化的奇妙结构,其中"漏"带来幽深感和神秘感,而"透"则具有通明感和空灵感。

Standards for a good rockery stone cherished by the Chinese intellectuals can be summarized as *shou* (being slim), *zhou* (having wrinkled textures), *lou* (having small holes), *tou* (having large and permeable holes). Slimness of stone is praised because since the Song Dynasty (960-1279) there has been a love of thinness among ancient Chinese. Wrinkles on the stone are formed after thousands-of-years' weathering and crustal movements. Therefore, they can express a sense of history. Small and large holes in different places can complicate the structure of the stone. In Chinese intellectuals' opinion, *lou* gives people a sense of mystery while *tou* gives a sense of transparency.

灵璧石
Lingbi Stone

- **苏州网师园冷泉亭内的灵璧石**

  灵璧石,产于安徽省灵璧县境内。清朝乾隆皇帝曾御封灵璧石为"天下第一石"。灵璧石最大的特点是叩之能发出优美的声音,在商周时代就曾被用于制作石磬。灵璧石又有奇石的形、质、色、纹等特质,在古代就是著名的观赏石。

  A *Lingbi* Stone in *Lengquan* Pavilion of the Master-of-nets Garden, Suzhou

  Lingbi Stone was produced in Lingbi County, Anhui province. Emperor Qianlong in the Qing Dynasty once called it "the Best Stone in the World". *Lingbi* Stone is best known for its mellifluous sound when knocked, and has been made into stone resonators as early as the Shang and the Zhou dynasties. Besides its sound, it is also cherished for its special shape, texture, color and veins. As a result, *Lingbi* Stone has been known and appreciated since ancient times in China.

房山石
Fangshan Stone

- 颐和园乐寿堂前的房山石"青芝岫"

房山石是北方园林叠石的主要用材之一，又名"北太湖石""土太湖石"，产于北京房山大灰厂一带的山上，河北、河南、山东泰山等地也有出产。房山石产于土中，因被红土渍埋而呈土红色或赤黄、淡黄色，质地坚韧，沉实浑厚，表面也有涡、沟、洞，但多为密集的小孔。青芝岫放置于颐和园乐寿堂前院内，石质是房山石，产自北京房山，长8米、宽2米、高4米，重约二十几吨，形似灵芝，是中国最大的园林观赏石。

A *Fangshan* Stone Called *Qingzhixiu* before the Hall of Joyful Longevity in the Summer Palace

Fangshan Stone, known as North *Taihu* Stone and *Tu Taihu* Stone as well, is the main stone used for piling in gardens in North China. It is produced in mountains near *Da Huichang*, Fangshan District, Beijing and some other provinces like Hebei, Henan and Shandong. Buried in soil, *Fangshan* Stone therefore is brown or light brown. It is hard and solid, full of whorls, small grooves and holes. *Qingzhixiu*, coming from Fangshan, Beijing, is a ganoderma-shaped big stone which is 3 m in length, 2 m in width, 4 m in leight and over 20 t in weight. It is the largest ornamental stone in the Chinese Garden.

英石
Ying Stone

- **案头清供英石"龙马拂波"**

  英石产于广东省英德县，是上乘的园林用石，颜色呈淡青灰，间有白色纹路，轮廓变化突兀，表面嶙峋，皱纹深密，精巧多姿，质地坚而脆，手指叩弹有共鸣声。岭南园林中常见英石，多用于立峰。

  A *Ying* Stone Called *Longmafubo* Placed on the Desk

  Ying Stone, produced in Yingde County, Guangdong Province, is a high-quality gardening stone in China. *Ying* Stone is light blue, striated with white lines. It changes abruptly in contours and shapes and wrinkles on its surface are deep and dense. Besides, it is hard, echoing pleasantly when knocked. *Ying* Stone is often used in gardens in Lingnan district, and is usually placed upright.

黄石
Yellow Stone

- **上海豫园大假山**

  黄石是古典园林叠山最常用的石料之一，以产于江苏省常州市的品种最佳。黄石那斧劈般的解理面具有苍劲、古拙的阳刚之美，常用于堆叠大型假山、拼峰或散置，极少独峰特置，在江南园林中的应用仅次于太湖石。上海豫园仰山堂前的大假山，依水而筑，是江南现存明代用黄石叠置的第一大山，出自叠山名匠张南阳之手。

  The Big Rockery in *Yuyuan* Garden, Shanghai

  Yellow Stone is a common piling stone in the Chinese gardens. Of all Yellow Stones, those produced in Changzhou, Jiangsu Province, are the best. The cleavage plane of Yellow Stone is rough and uneven, like being cut by axes, giving a sense of vigorousness and simplicity. Yellow Stones are usually piled together, or piled with other kinds of stones together, into a high rockery. Rarely is a stone a rockery. In gardens on the Yangtze Delta, it is frequently used, only second to *Taihu* Stone. The big rockery before *Yangshan Tang* in Shanghai *Yuyuan* gardens is the biggest in extant Chinese Garden, and is constructed by Zhang Nanyang, a famous stone pilling maker.

青石
Bluestone

- 北京中山公园青云片

青石在北京园林中用得很多，大部分产于北京西郊红山口。青石形体多呈片状，以横纹取胜，因而又有"青云片"之称。在北京中山公园来今雨轩南侧，有一块青石点景"青云片"。石高3米，长3.2米，周长7米。石色发青，玲珑剔透，姿态优美。这块青石原放在圆明园时赏斋院内，石上刻有清乾隆帝亲题的"青云片"三字，圆明园被焚毁后移至今中山公园内。

A Bluestone Called *Qingyunpian* in Beijing *Zhongshan* Park

Bluestone, produced in Hongshankou in west Beijing, is frequently used in gardens in Beijing. Bluestones are mainly slates, known for the cross striations. Therefore, it is also called *qingyunpian* (a blue cloud-like slate). In south of *Laijinyu* Veranda of Beijing *Zhongshan* Park, there is a *qingyunpian* which is 3 m high, 3.2 m long and has 7 m in its circumference. The stone looks blue and green. It is primarily placed in *Yuanmingyuan* Garden, with its name *qingyunpian* caved on it by Emperor Qianlong of the Qing Dynasty. After *Yuanmingyuan* Garden was destroyed by the foreign invaders in the late Qing Dynasty, the stone was moved to *Zhongshan* Park.

宣石
Xuan Stone

- **宣石**

  宣石产于安徽省宣城市，颜色洁白，形体浑厚，石上没有挺直的棱角，也没有环洞变化。由于宣石产在山中，在红土中长期积渍，一般带有赤黄色，须经过刷洗才见石质。

  ***Xuan* Stone**

  Xuan Stone, produced in Xuancheng, Anhui Province, is white and hard, with no upright edges or whorls and holes. Buried in red soil on the mountain, *Xuan* Stone is usually brown before being cleaned.

掇山

　　掇山就是将选好的山石叠放成需要的假山制式，以制造出深幽、峭拔、灵秀的特色。园林设计中，假山为空间的重要组成部分，能划分园区、隔离视线，同时也可作为主景、远景及背景。

庭山

　　庭山是指修建在前庭院中的叠石。堂前叠石多作观赏，取崇山峻岭之意，更多是将姿态美观的树与山石相搭配。修建在厅堂前的叠石叫"厅山"；修建在书房前的叠石叫"书房山"。

Hill Making

Hill making refers to the piling of rockeries with stones. It is an important part in gardening, because rockeries may help separate spaces and be used as screens. They could be main scenery, distant scenery, or background scenery.

Tingshan

Tingshan refers to piled stones in gardens before the houses. Those stones, decorated with beautiful trees, are for ornamentation and appreciation. The piled stones in front of halls are called *Tingshan* (meaning "mountains in front of halls" in Chinese) and those in front of the studying room are called *Shufangshan* (meaning "mountains in front of the studying rooms" in Chinese).

- **北京北海公园快雪堂前的云起石**

  云起石为两块高大的房山石，挺拔峻峭，巍峨壮观，形状犹如云头对起，故称"云起石"。石的正面刻有乾隆皇帝御题"云起"二字，背面刻有乾隆皇帝御题《云起峰歌》。

  Yunqi Stone in front of the Kuaixue Hall in Beihai Park, Beijing

  Yunqi Stone is formed by two big *Fangshan* Stones. It is called *Yunqi* (meaning "surging clouds" in Chinese) because it looks like two sheets of surging clouds. In front of the stone there are two Chinese characters *yun* and *qi*; at the back is a poem called *Ode on the Yunqi Mountain*. All were written by Emperor Qianlong of the Qing Dynasty.

壁山

又名"峭壁山",多见于江南较小的庭院内,是在粉墙上嵌入山石,或依墙壁叠石,抑或在墙壁中嵌山石成石景、山景。以粉壁为背景,恰似一幅中国水墨画,特别是通过洞窗、洞门观赏,画意更浓。

Bishan

Bishan, also named *Qiaobishan*, commonly exists in small gardens on the Yangtze Delta. It refers to rockeries embedded in the wall or placed close to the wall. The stones and walls are delicately united like in a Chinese ink painting, especially when they are appreciated through windows or doors with holes.

- 苏州拙政园海棠春坞

苏州拙政园海棠春坞庭院的南面院墙上有用山石嵌成峭壁山一座,并种植海棠、慈孝竹,题名为"海棠春坞"。

Malus Spring Castle in the Humble Administrator's Garden, Suzhou

On the south wall of the garden where the Malus Spring Castle is there is a *qiaobishan*. The rockery is decorated with malus and bamboo trees, so it is named *Haitangchunwu*.

楼山

楼山，指的是以叠山作为楼阁的基础，或叠石成石涧、石屋，在上面建造楼阁。游人可登山入楼远眺，山石建筑浑然一体，洞和阁相连，增加了变化，显得生动自然。叠石还可做成自然踏垛，作为楼阁的室外楼梯，称"云梯"。

Loushan

Loushan in Chinese means pavilion rockery because a pavilion is constructed on it. Tourists can climb the rockery into the pavilion and look at distant sceneries. *Loushan* has united stones and buildings together in a interesting and natural way. Stones piled as the stairs of the pavilion are called Cloud Stairs.

- 苏州狮子林卧云室

卧云室为假山环抱中的方形楼阁，共有两层，造型十分奇特。楼阁周围被重叠的山石层层包裹，人在其中，仿佛处于石壁重重的山坳之中。

The *Woyun* Pavilion in the Lion Grove Garden, Suzhou

The *Woyun* Pavilion is a square pavilion constructed on rockeries. The two-floored pavilion with a very unusual appearance is surrounded by rockeries. Tourists in it feel like in a mountain col.

池山

池山，就是在水中叠石成山，可同时有峰、峦、洞、穴、涧、谷、石壁之美，颇具深山幽谷、咫尺山林的意境。

Chishan

Chishan in Chinese means pool rockery because the stones are piled in the pool water. *Chishan* has united rockeries and water together and thus produces a sense of seclusion and peacefulness.

- 苏州拙政园中的池山
The *Chishan* in the Humble Administrator's Garden, Suzhou

山洞

洞，是山体内部中空的空间，通常设有石台、石凳，供人歇息。洞口要与整个山体浑然一体，洞内空间也要随着山势力求自然，使游人犹如进入天然山洞。

Rockery Caves

Usually, stone tables and stone benches are placed in the rockery caves. The mouth of the cave and the ups and downs in the cave should be made naturally as in a real one.

- 苏州沧浪亭假山洞
Rockery Caves of *Canglang* Pavilion in Suzhou

磴道

园林中山石的设置讲究顺应地形，往往利用山石堆成各种形式的磴道，可以将游人引入奇妙的胜地，形成一种深邃神秘的意境。

Stone Path

The stone paths in the Chinese Garden should be paved according to the shape of the land. These stone paths could lead the tourists to places unexpected and thus produce a sense of mystery.

- 北海琼华岛假山的石磴道
The Stone Path on the *Qionghua* Islet, *Beihai* Park

置石

置石又名"点石"，是选用少量山石在园林中巧妙地点缀、布置成景的手法。置石的位置常选在庭院、池塘边、路边、墙角、土山等处。常用的布局方式有特置、对置、散置、群置等。

Stone layout

Stone layout, known as *dianshi* as well, is a technique for arranging stones. In the technique, a small amount of stones are piled and scattered in the garden. They are usually placed at a pool, on the roadside, in the corner of a wall, or at an earth-piled mound. The usual ways of layout are special layout, coordinate layout, scattered layout, and crowded layout.

- 苏州留园岫云峰

岫云峰是留园中特置的峰石之一，石高5米，在冠云峰的后边右侧，石形甚美，春季时，岫云峰上有青藤、绿藤缠绕，更具情趣。

Xiuyun Rockery in the Lingering Garden, Suzhou

Xiuyun Rockery is a 5-meter rockery in the Lingering Garden. The shape on its back right side is especially attractive. In spring when ivy is growing, the rockery looks even more beautiful.

- 北京颐和园中对置的山石

Two Oppositive Rockeries in the Summer Palace

- 无锡寄畅园中散置的假山石
Scattered Stones in the *Jichang* Garden, Wuxi

- 苏州留园中群置的假山石
Grouped Stones in the Lingering Garden, Suzhou

> 水景的设计

　　水和山都是中国园林重要的构景要素，有山必有水，青山绿水构成园林的灵魂。在中国园林中，各种水体是自然界中河、湖、溪、涧、泉、瀑的集中体现。对于水体景观的营造多用桥、堤、岛、汀步分隔水面，或以亭、台、榭、廊划分水面，并以莲荷、菖蒲等花草点缀水面。

湖

　　园林中的湖多是在天然水体的基础上略加人工开挖而成，周围常有山脉起伏，形成湖光山色的典型风景。

　　在园林中，湖一般作为中心景区，沿湖岸多布置道路、建筑、山石、花木，形成种种景色。浩瀚的湖面常用堤、岛、桥等加以分隔，成为

> Design of Water Scenery

Water and mountains are the important components in the landscaping of Chinese gardens. Mountains are always arranged together with water. Green hills and clean water form the spirit of gardens. In Chinese gardens, water is the main embodiment of rivers, lakes, brooks and streams, springs and waterfalls in nature. To construct a water scenery, people often use bridges, embankments, islets and stepping stones to separate the water surface, or build pavilions, terraces, pavilions on terrace, and corridors to divide the water, and plant lotus, calamus and other flowers to decorate the water surface.

Lake

Lakes in gardens are often artificially dug based on natural water. Often there are rolling hills around the lakes which form a typical landscape of lakes and mountains. Lake is often set as the central scenic spot

有对比、有层次的空间水域。由于湖岸线长，曲折蜿蜒转且具自然变化，用树木、建筑适当点缀，可使湖岸景致呈现优美多变之势。

- **颐和园昆明湖**

  颐和园里的昆明湖面积达2.2平方千米，粼粼的湖水、蜿蜒的长堤、错落的岛屿以及湖畔的各式建筑，组成了颐和园以水为主体的风景。

  Kunming Lake in the Summer Palace

  Kunming Lake in Summer Palace covers a total area of 2.2 square kilometers. The sparkling lake water, the long and winding embankment, the scattered islets and the various buildings by the lakeside form the water landscape of the Summer Palace.

of the garden. Along the embankment roads, buildings, hills and stones, flowers and trees are often arranged to form various landscapes. Vast expanse of lake is often separated by embankments, islets and bridges to form contrastive water space with different levels. The long and winding lake shoreline, natural changes, and the appropriate decoration of plants and buildings make the various changes of the beautiful scenery along the lake embankment.

池

池主要用于构景，常见于规模不大的园林或大型园林中的局部景区。在水面较大的池中往往设岛，或用桥、廊划分水面，使空间贯通，隔而不分；狭长的水池，纵向景深，水湾萦绕，更觉幽曲不尽。池周山石、花木、亭榭映入水中，与天光云影、碧波水色相融，更增添了园景生气。

- 苏州网师园彩霞池
  Caixia Pool (Rosy Cloud Pool) in Garden of Master-of-nets in Suzhou

Pool

Pools are mainly used in landscaping, which is often seen in small gardens or the local landscape areas in large-size gardens. For large pools islets, bridges or corridors are often established to separate the water surface and at the same time connect different spaces. For long and narrow pools, the vertical scenery is deep and rich and the bays wind around, which gives an impression of endless tranquility. The reflections of hills and stones, trees and flowers, and pavilions around the pool, together with the reflections of the sky and clouds bring vim and vitality to the landscape.

河流

园林中的河常借助自然水系，一般都富于动感而且具有天然的旖旎风光。如扬州瘦西湖凭借保障河水域的天然地势，营造出若干优美的园林景观。

River

Rivers in gardens often make use of natural water, which are usually full of dynamics and the charming natural scenery. Such as the Slim West Lake in Yangzhou, with the help of the natural terrain of *Baozhang* River, creates several beautiful garden landscapes.

● 北京颐和园苏州河
Suzhou River in the Summer Palace, Beijing

溪涧

园林中设置溪涧，大都模仿自然山涧小溪，水流有分有合，水面有宽有窄，有的在岩洞间穿行，有的在亭榭间环绕。

Brooks and Streams

Brooks and streams in gardens often imitate brooks and streams in nature. The water separates and combines now and then; the water surface is vast or narrow; some winds through caverns and others flows around between pavilions.

- 溪涧
  Brooks and Streams

瀑布

在园林中设置人造瀑布，只能取其意境。常利用山石的高差来引水流动；也有以建筑屋檐流下的雨水为水源，在假山上作水口，仅暴雨时才能见到瀑布。水瀑可呈帘状、柱状或线状，有直落、飞落或分段迭落各种形式，瀑下常有潭或池。

Waterfall

To set man-made waterfalls in gardens is to imitate the images of them only. The altitude difference of mountains and rocks is employed to drive the water to flow. Sometimes rains on the roof of the buildings are taken as the water source and artificial mountains as water outlet. In this way waterfall can only be seen when heavy rain falls. Waterfalls can take on shapes like curtains, columns, or lines, straightly falling down or flying down or intermittently falling down. Below waterfalls there often are deep ponds or pools.

- **苏州狮子林中的飞瀑**

  在狮子林中，有一处苏州古典园林中独有的景观，即飞流直下的瀑布。瀑布以三跌倾泻而下，在出水口还夹有两峰石，一跌处跨石梁。瀑布的旁边建有一座飞瀑亭。亭南有瀑布流泉自山顶而下，景观优美。人在亭中可以观赏瀑布，更能倾听瀑布之声。

  Flying Waterfall in the Lion Grove Garden in Suzhou

  In the Lion Grove Garden there is a unique landscape in Suzhou classical gardens, that is, the flying waterfall. The waterfall pours down in three sections. At the outlet there is a rock with two peaks and the first section of the waterfall stretches across a stone beam. Next to the waterfall is the Pavilion of Flying Waterfall. In the south of the pavilion one can see the waterfall pouring down from the peak of the mountain, which is really a beautiful scene. In the pavilion one cannot only watch the waterfall but also listen to the sound of it.

泉

园林中的泉，景从境出。若山泉，便引导泉脉，布置成深潭、池沼，任其涌流。若平地涌泉，便于庭院中做池，围以壁栏，筑以亭盖，池底铺粗砂，池壁上做出水口，水口用石雕兽头加以装饰。

Spring

Springs in gardens breed the scenery. If it's a mountain spring, the source of the spring will be guided into courtyards to build a pool. The pool will be enclosed with railings, and a pavilion will be built. The bottom of the pool will be paved with sands, and outlets which are decorated with stone-carved beast heads will be built on the pool wall.

• 山东济南趵突泉 (图片提供：图虫创意)
Baotu Spring in Jinan, Shandong Province

岛

中国古典园林有"一池三山"的传统造景手法，即指水中设岛。岛划分了水面空间，增加了风景层次，可观可游。岛的四周视野开阔，是眺望风景的观赏点，岛又处于四周视线的焦点，也是被观赏的对象。

Islet

In the construction of Chinese classical gardens there is a traditional approach named "one pool and three mountains" which refers to the establishment of islets in water. Islet separates the space of the water surface and enriches the landscape levels, which makes the island a good place to watch and tour. Islet is a scenic viewing point because it provides an open and broad view of the landscapes around; island is an object of viewing because it is the focus of attentions around it.

- 北京颐和园南湖岛

南湖岛位于昆明湖最大的东水域中心，与万寿山隔湖遥遥相对。岛的北、西、南三面都有很好的观景条件，从涵虚堂往北、往西眺望，近处的万寿山、西堤，远处的玉泉山、西山和烟波浩渺的湖面统统纳入眼帘，构成一幅将近2000米长的风景画面。南湖岛与东岸之间，则以多彩多姿的十七孔长桥相连接。

The South Lake Islet in the Summer Palace, Beijing

The South Lake Islet is situated at the center of eastern water of *Kunming* Lake, facing the Longevity Hill across the lake. The north, west and south sides of the islet have very good viewing conditions. Viewed from *Hanxu* Hall northward and westward, the Longevity Hill, the West Bank not far away, and the *Yuquan* Hill, the Western Hills and the wide expanse of misty lake surface come into view and form a picture of landscapes near 2,000 meters long. The South Lake Islet and the East Bank are connected by the colorful and beautiful Seventeen-arch Bridge.

堤

堤也是中国园林中重要的水景形式。沿水边设置用于拦水、防水的建筑物叫一面临水堤；也有堆在湖水中起分割湖面作用的两面堤，如西湖中的苏堤、白堤。长堤属于野外风景，堤上植柳栽桃，建桥、亭廊、花架，布桌椅儿凳，造成局部景观的丰富多样和序列构图的节奏变化。

Embankment

Embankment is also an important feature of water scenery of Chinese gardens. Bank established to bar and prevent water is the one-side waterside embankment; there are also double-side embankments in the lake to divide the surface of the lake, such as the *Su* Causeway and the *Bai* Causeway in the West Lake. Long embankment is a landscape in the wild. On the embankment willow trees, peach trees are planted, bridges, pavilions and corridors, flower pergolas are built, and desks, chairs, and benches are arranged, which result in the richness of the local landscape and the rhythmic changes of the design of the landscape.

- **杭州西湖苏堤**

杭州西湖总面积约5.6平方千米，湖中构筑了纵贯南北的苏堤和横贯东西的白堤，把全湖分割成五个部分。苏堤上设有六座形制不同的拱形石桥，使得西湖之水隔而不断，富有情趣。

The *Su* Causeway in Hangzhou West Lake

Hangzhou West Lake covers a total expanse of 5.6 square kilometers. The *Su* Causeway from south to north and the *Bai* (White) Causeway from east to west in the lake separate the entire lake into five sections. On the *Su* Causeway there are six stone arch bridges with different shapes, separating and also connecting the West Lake water, which is full of fun.

汀石

汀石又名"汀步""步石"，是仿自然界溪谷水景，在池窄水浅处点置若干石块代替桥，使水景更富自然形态和野趣。汀石选较大的外形不整而石面较平的石块，砌在水底，露出水面的部分可大小高低参差错落，两块相邻的步石间距大约为一步的距离。

Tingshi (Stepping Stone)

Tingshi is also known as *Ting* Step or Stepping Stone. It is an imitation of water scenery of valleys in nature by placing several stones to replace bridges in the narrow and shallow places of the pool, which makes the water scenery much more natural and full of rustic charm. Large stones of irregular shape and flat stone surface will be selected as *Tingshi* to be built on the bottom of pools. The parts of these stones exposed on the surface can be of different sizes and heights, and the distance between two stones is about one step.

- 汀石
  Tingshi (Stepping Stone)

> 花木的点缀

中国地域辽阔，气候类型多样，植物种类也非常丰富，常用来种植在园林中作为点缀。园林中的花木，从观赏的角度讲，可以分为观花类、观果类、观叶类、松树类、藤蔓类、竹类以及水生植物等几个种类。

> Ornament of Plants

China is a country with a vast territory, various climates and rich plant species. Plants are often used as decoration in gardens. Trees and flowers in gardens, from the perspective of aesthetic enjoyment, can be divided into several categories: flowering plant, fruit plants, foliage plants, pine trees, vines, bamboos, aquatic plants, and so on.

- **南京瞻园池岸边的爬山虎**

  爬山虎一般种植在园林中高大的建筑物或山石旁，可以将墙壁和山石完全覆盖，形成大面积的绿色景观。

  Parthenocissus Tricuspidata by the *Zhanyuan* Pool in Nanjing

  Parthenocissus tricuspidata is usually planted by the side of high buildings or rocks in the garden, which can completely cover the walls and rocks, forming a large area of green landscape.

• 北京北海岸边的垂柳

观叶类植物以植物叶子的形状与姿态为观赏对象。柳树枝叶修长，姿态柔美纤弱，尤其适宜栽种在水边，显得翩翩生姿，风情万种。

Weeping Willows on the Embankment of *Beihai* in Beijing

Foliage plants take the shapes and postures of the leaves as the object of appreciation. The twigs and leaves of weeping willows are slender, and the postures are soft and delicate. Weeping willows are especially suitable to be planted at the waterside, which are elegant and charming.

• 果实累累的石榴树

观果类植物有枇杷、橘子、无花果、石榴等。植物枝头挂满累累硕果，有的还可以品尝，令人感到生命的充实。

Pomegranate Trees Heavy with Fruits

Fruit plants include loquat, orange, fig, and pomegranate, etc. When trees are heavy with fruits, some of which are edible, people can feel the fullness of life.

• 高大的银杏树

为了营造清幽的氛围，园林中常常栽种一些枝繁叶茂的树木，如梧桐、香樟、银杏、合欢、皂荚、槐树等。这类树木高大粗壮，枝叶茂密，以巨大的树冠遮蔽出成片的浓荫。

A Large Gingko Tree

In order to create a tranquil atmosphere, some lush trees such as Chinese Parasol, camphor trees, gingko trees, silk trees, Chinese honey locust and Chinese scholar trees are often planted in gardens. These trees are very large trees with strong branches and heavy leaves, and also a huge crown covering an expanse of shade.

• 苏州退思园内的紫藤

藤蔓类攀缘植物依附缠绕在墙垣或山石上，形成一种牵牵连连的纠缠的美感。紫藤性喜攀援，生长形态不受约束，可以为园林景观填补空白，增加生气。

Chinese Wisteria in the Retreat and Reflection Garden in Suzhou

Vines creep along or attach to the wall and rocks, forming a sense of beauty of entanglement. Chinese wisteria likes climbing and its growing is unfettered, which fills the blank of garden landscapes and brings vim and vitality to the garden.

• 天津曹家花园芦苇茂密的水面
Water with Dense Reed in the *Cao* Family Garden in Tianjin

• 北京北海团城内的古松"遮阴侯"

位于北海公园团城内承光殿东侧的松树，树形古拙，树冠庞大，好似撑开的巨伞。据说一年夏天，清乾隆皇帝游北海，到团城适值正午，天气炎热。乾隆帝坐在这棵油松树荫下休息，清风拂过，暑汗全消，乾隆帝非常高兴，封这棵油松为"遮阴侯"。

The Old Pine Sunshade Duck in the Round Castle in *Beihai* Park, Beijing

The pine tree on the east side of *Chengguang* Hall in the Round Castle in *Beihai* Park has a very old shape and a huge crown which looks like a huge opened umbrella. It is said that one summer day emperor Qianlong of the Qing Dynasty took a trip to *Beihai*. When he reached the Round Castle, it was just high noon and the weather was very hot. The emperor took a break in the shade under this Chinese pine tree. When the breeze blew, the emperor felt very cool and comfortable. Happily the Emperor conferred the title Sunshade Duke on the pine tree.

• 苏州沧浪亭的竹间小径

竹子是中国古代文人青睐的一种植物，它不仅修长俊秀，风度翩翩，而且自古就象征着高风亮节和君子之风。中国园林内常植有紫竹、斑竹、寿星竹、观音竹、石竹等竹类植物。而竹子与山石相配，则是园林中最经典、最传统的组合之一。

Bamboo Path in the *Canglang* Pavilion in Suzhou

Bamboo is a very popular plant among Chinese ancient literati. It is not only slender and graceful but also symbolizes exemplary conduct and nobility of character and the manner of gentleman since ancient times. In Chinese gardens bamboo plants like black bamboo, mottled bamboo, Buddha bamboo, Bambusa multiplex and Dianthus chinensis are often planted. The match of bamboo and rock is one of the most classic and traditional combinations in garden landscaping.

• 无锡蠡园的荷花

在池中种植荷花，是古典园林的传统，因此园林中心的水池常被称为荷花池。

Lotus in *Liyuan* Graden in Suzhou

It is a tradition to plant lotus in pools in classical gardens. Therefore, pools in gardens are often called the Lotus Pool.

• 海棠

观花类植物，常见的如梅花、菊花、桃花、桂花、山茶花、迎春花、海棠花、牡丹、芍药、丁香花、杜鹃花等。这些植物开花时都色彩绚丽、姿态婀娜、芳香沁人，经常成片种植在厅堂前的空地上，供人观赏。海棠花经常种植在院落中，且与象征富贵的牡丹花一起成片种植，寓意"满堂富贵"。

Chinese Flowering Crab-apple(Crab-apple)

Common flowering plants are Chinese plum, chrysanthemum, peach, osmanthus, camellia, winter jasmine, Chinese flowering crab-apple, peony, Chinese herbaceous peony, lilac, azalea, etc. The flowers of these plants are colorful and in graceful shapes and have pleasant fragrances, which are often grown in a large scale in the clearing in front of halls for the enjoyment of tourists. Chinese flowering crab-apple, usually planted in courtyards, is often grown with peony which symbolizes wealth in a large scale, implicating that the family is replete with prosperity and honors.

中国园林中花卉的种植方式
Planting Methods of Flowers in Chinese Gardens

花台
Flower Terrace

花畦
Flower Bed

• **苏州环秀山庄的花台**

规则式花台用砖石砌筑，有圆形、半圆、方形、长方形、多边形及海棠、梅花等形状，多用于庭院中，与规则的空间形式相呼应。自然式花台多用黄石或湖石叠筑，平面、立面都采用不规则构图，布置于厅前、屋后、轩旁廊侧、山脚池畔等。花台内配置花草树木，尤讲究造型姿态，辅以石峰石笋，构成天然图画。

Flower Stand in *Huanxiu* Mountain Villa in Suzhou

Regular flower stands are built with bricks and stones which are round, semi-circular, square, rectangular and polygonal, or in the shape of flowers of Chinese crab-apple or Chinese plum and so on. Flower stands are often used in courtyards, echoing with the regular space arrangement of the garden. Natural flower stands, mostly built with stacked Yellow Stones or Lake Stones, the plane and the façade of which adopt irregular composition, are placed in the front of hall, behind houses or on both sides of rooms and corridors, or at hill foots and beside the pools. Flowers, trees and grasses will be planted in flower stands, with special attention paid to the design and postures of them. Stone peaks and stalagmites will be added into flower stands to constitute a natural picture.

• **苏州拙政园花畦**

Flower Bed in the Humble Administrator's Garden in Suzhou

花畦又称"花池子"，一般呈方形、长方形、条形，用细竹编成矮篱或以砖瓦砌边，也可植草作为畦边。畦中丛植一种花卉或群植多种花卉。多布置于路径两侧、廊前，以条形为多，也有砖瓦砌成各种形状的小型花池，连续排列。

Flower bed is also called flower pool, which is usually square, rectangular or in strip-style. It is enclosed with short bamboo fences or walls with bricks, or grass. In the flower bed a single kind of flower or several kinds of flowers are planted together. Flower beds are often built in continuous rows on both sides of roads or in front of corridors, which are often in the shape of strips, or small flower pools of various shapes.

花圃
Flower Plot

- 花圃
古典园林中常修建一个花圃，专门种植几种花木而成景，花圃中所种植的花木大多不是名贵的花木。
Flower Plot
Flower plots are often built in classical gardens. Several kinds of trees and flowers are specially planted to form a landscape, most of which are not rare trees and flowers.

中国园林中的建筑
Architectures in Chinese Gardens

　　建筑是中国古典园林造景的主要手段之一，常见的园林建筑有厅、堂、轩、馆、楼、阁、亭、榭、廊、舫等。园林建筑以其独特的造型轮廓与山、水、花木共同组成风景画面，在局部景区中往往成为构图的中心和主体，同时还是游览休息场所、观景的地点。中国园林建筑重视造型轮廓的轻巧玲珑，飞檐飞翘，曲线优美，体量尺度相宜，与周围景物和谐。在装饰色彩上讲究精巧细腻、雅朴自然。

Architectures are one of the primary means of landscaping in classical Chinese gardens. Common garden architectures include, halls, bowers, pavilions, storied buildings, towers, pavilions on terrace, corridors, boat-like structures, etc. With the uniqueness of their profiles, garden architectures constitute scenic landscapes together with mountains, water and plants. While they are usually made as the focus and subject of the picture, garden architectures also provide space for rest and sightseeing. Chinese garden architectures pay much attention to the style: its shape must be exquisite, its cornice curvy, its outline graceful, and its scale appropriate when compared with the other objects in the scene. The colors that decorate these garden architectures are particularly delicate and natural.

> 园门建筑

园门，又叫门楼，是园林的入口。有的园林除设置大园门以外，还在园林内的小园林中设置小园门。皇家园林都采取宫殿大门的形式，以体现皇权的赫赫威势；而私家园林的园门，因园主身份的不同而显得丰富多彩。

> Garden Gates

Garden gates (*yuan men*), also known as gate houses (*men lou*), are the entrances of the gardens. Some gardens, apart from a major garden gate, also have minor garden gates for the small gardens within. The garden gates of the royal gardens are all in the form of palace gates, in order to express the authoritativeness of imperial power. However, those of the private gardens differ because of the variety of their owners' social status and position.

Screen Wall

Screen wall (*yingbi*, aka. *zhaobi*) is a wall which stands by itself inside or outside of the gate. Usually, a screen

- 北京颐和园北宫门
The North Gate (*Bei Gongmen*) of the Summer Palace, Beijing

影壁

影壁也称"照壁",是一面独立的墙壁,立于中国古代院落大门内外,常和大门外左右的牌楼一起组成门前广场,是整组建筑的"序幕"和先导,同时也有象征权势的作用。位于大门之内的影壁,叫"内影壁",作用主要是挡住外人的视线,使之不能对院内一览无余。影壁虽然是一面墙壁,但由于设计巧妙,施工精细,往往能起到烘云托月、画龙点睛的作用。

wall and the archways on both sides of the main gate constitute a plaza at the front door. The screen wall serves as the prelude and introduction to the whole set of constructions and at the same time, symbolizes power and influence. The screen walls inside the gate are internal screen walls (*nei yingbi*), whose main function is to block out the passersbys'sight so that they can't cover all at a glance. A wall it is, but because of the clever design and fine construction, the screen wall is often the foil or the finishing touches that highlights the whole picture.

- **北京北海九龙壁**

  北京北海公园北岸有一座琉璃九龙壁,原为佛教庙宇——大圆镜智宝殿山门前的影壁,高6.65米,壁面双面各有用琉璃烧制的红、黄、蓝、白、青、绿、紫七色蟠龙九条,戏珠于波涛云雾之中,形态各异,栩栩如生。

  Nine-dragon Wall in *Beihai* Park, Beijing

  On the north bank of *Beihai* Park, there is a nine-dragon wall made of colored glaze. This wall was originally the screen wall outside of the mountain gate (*shan men*) of a Buddhist temple—the Temple of Mirror-like Wisdom (*Dayuanjingzhi Baodian*). The Nine-dragon Wall is 6.65 meters in height. Colored-glaze of red, yellow, blue, white, cyan and purple are placed in the image of nine dragons on each side. The dragons are playing with pearl in the sea of clouds; each of them varies in shape and is seemingly alive.

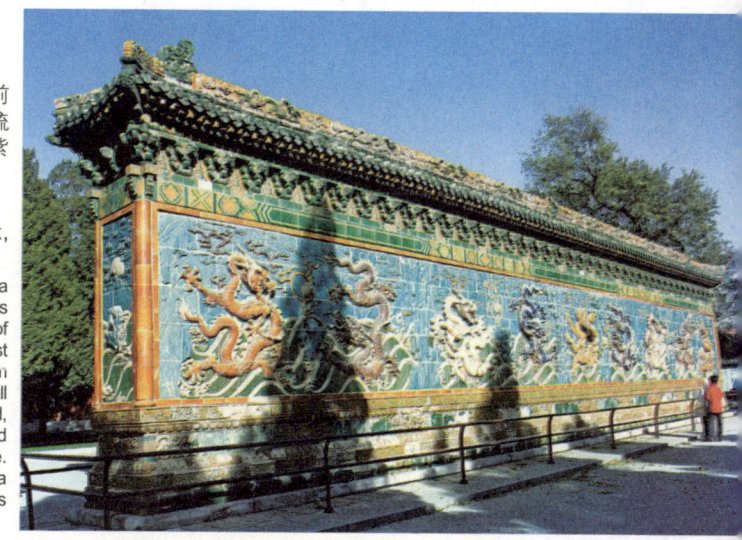

牌坊

　　牌坊又称"牌楼",无依无傍,巍然耸立,是一种形象华丽的建筑物,往往作为入口的标志,也具有大门的功能。它以丰富的造型和精美的装饰、绚丽的色彩令人注目,成为具有中国风格的民族建筑之一。牌楼按照规模大小,有一间二柱、三间四柱、五间六柱等几种,其中以三间四柱式最为常见。

Memorial Archway

Memorial archway (or memorial gateway) is usually a lofty architecture that stands majestically. It signals the entrance and functions as a gate. With rich styling, fine decoration and brilliant colors, memorial archways stand out as one of the most typical Chinese style national architectures. According to the size, memorial archways are classified into different groups: two-column, four-column, six-column, etc, among which the most popular style is the four-column style.

- 北京颐和园知鱼桥牌坊

知鱼桥牌坊位于颐和园万寿山东麓的谐趣园内,据说是京城最小的石牌坊。

The Memorial Archway on Knowing-the-fishing-bridge (*Zhiyu Qiao*), the Summer Palace, Beijing

This memorial archway is in the Garden of Harmonious Interests (*Xiequ Yuan*), which is situated at the bottom of the eastern side of the Longevity Hill (*Wanshou Shan*). It is said to be the smallest stone memorial archway in Beijing.

• 北京颐和园东宫门外木牌楼

颐和园东宫门外的木牌楼，正面额上写着"涵虚"，影射前面的水景；背面额上写着"罨秀"，暗指背面的山景。

The Wooden Memorial Archway outside the East Palace Gate of the Summer Palace, Beijing
The tablet on the front says *hanxu*, which echos with the water feature in front of it; the tablet on the back says *yanxiu*, which suggests the mountains behind it.

垂花门

垂花门是中国古建筑中常见的一种门，因其檐柱不落地而垂吊在屋檐下，其下有一垂珠，通常木雕彩绘为花瓣的形式，故称为"垂花门"。

垂花门的装饰性极强，它的各个突出部位几乎都有十分讲究的雕饰。从外边看，垂花门像一座华丽

Hanging Flower Gate

Hanging flower gate (aka. festooned gate) is frequently seen in ancient Chinese buildings. It gets its name from the flower carvings beneath the posts. The posts are not connected to the ground; instead, they hang under the eave. The carvings are usually made on the wood and are in the shape of petal.

Hanging flower gate is highly

的砖木结构门楼；而从院内看，则似一座亭榭建筑的方形小屋。在园林中，垂花门一般作为园中之园的大门，或用于居住并兼具游赏性小园的正门或侧门和后门、垣墙上的随墙门、游廊的廊罩，起分隔园区、隔景、障景等作用，建筑形式灵活多样。

- 北京故宫宁寿宫花园遂初堂垂花门
The Hanging Flower Gate in the Hall of Wish Fulfillment (*Suichu Tang*), Palace of Tranquil Longevity (*Ningshou Gong*), the Forbidden City, Beijing

decorative. Almost every protruding part of the gate has dainty carving-and-engraving decorations. When viewed from the outside, the hanging flower gate looks like a gorgeous gate house made of wood and bricks. And while viewed from the inside, it looks like a small square pavilion. In gardens, hanging flower gates usually exist as the gates to the gardens within gardens, the front door of small gardens for living and sight-seeing, side doors, back doors, *suiqiang* doors (doors built in the wall), and doors for corridors. This style of gate is used to separate different garden parts and scenes, or to block certain views. And so their appearances are more flexible and various.

砖雕门楼

砖雕门楼多见于江南园林中，尤以徽式园林中更为多见。它作为建筑物的入口标志，采用考究的雕刻装饰手法，构成各种不同的造型，打破了平整的白粉墙面的单调之感，增添了古建筑的艺术美。砖雕的题材十分丰富，有取自吉祥图案，有植物纹样，有人物故事，构思精巧，技艺精湛。

Carved Brick Gatehouse

Carved brick gatehouses are mostly seen in gardens on the Yangtze Delta, especially *Hui*-style gardens. As the sign of entrance, carved brick gatehouses engage exquisite carvings to form various shapes. This erases the monotony of smooth whitewashed walls, and brings the aesthetic beauty of classical architectures. The themes of the brick carvings are full of variety— auspicious patterns, plants, and stories. Each of the carvings is created with ingenious design and crafted with great skills.

- 苏州网师园万卷堂"藻耀高翔"门楼

网师园主厅万卷堂前的砖雕门楼，高约6米，宽3.2米，厚1米，已有300余年的历史，以雕刻细致入微、风格秀丽而著称，享有"江南第一门楼"的盛誉。门楼上的砖雕运用平雕、浮雕、镂雕和透空雕等手法雕凿而成，人物栩栩如生，飞禽走兽和花卉图案形象逼真，为传统砖雕艺术中的精品。

Zaoyao Gaoxiang Gatehouse of Ten Thousand Volume Hall, Master-of-nets Garden, Suzhou

The carved brick gatehouse in front of the Ten Thousand Volume Hall (*Wanjuan Tang*) in the Master-of-Nets Garden (*Wangshi Yuan*) is about 6 meters high, 3.2 meters wide and 1 meter thick. With a history of more than three centuries, this gatehouse is famous for its fine carving and elegant appearance and enjoys a reputation as No.1 Gatehouse on the Yangtze Delta. With engagement of crafts like flat carving, relief carving, pierced carving, and hollow carving, the carved bricks on the gatehouse show vivid figures, animals and plants, making the gatehouse an exquisite example of traditional carved bricks.

> 宫殿建筑

宫殿，是中国古代帝王专用的居所或者供奉神佛的建筑，等级最高。对于园林来说，只有皇家园林

> Palaces

Palace, as the exclusive residence for the emperors in ancient China or the place used for worship of gods or Buddha, ranks the highest in all kinds of architectures. As

- **承德避暑山庄烟波致爽殿内部**
  烟波致爽殿是避暑山庄正宫区的后寝殿。大殿面阔七间，前后带廊，单檐卷棚歇山式灰瓦顶。大殿内部正中三间为厅，是皇帝接受后妃朝拜的地方，后方悬有"烟波致爽"匾。室内陈设丰富，富丽堂皇。

  The Inside of the Hall of Cool Mists and Ripples, Chengde Mountain Resort
  The Hall of Cool Mists and Ripples (*Yanbo Zhishuang Dian*) is the bedchamber area in the Front Palace of Chengde Mountain Resort. The main hall is seven-bay (literally room) wide, with colonnades in the front and on the back. It has a gray gable and hip roof with single eave and round ridge. The three bays in the middle of the main hall is the nave, where the imperial consorts paid respects to the emperor. At the back of the nave, there hangs a plaque that says *Yanbo Zhishuang*. The furnishing inside is rich and luxurious.

中才会有宫殿类建筑。宫殿类建筑形制高大，布局采用主殿居中，配殿分列两旁的对称形式，殿前有宽阔的庭园及广场衬托，洋溢着浓重的宫廷气息。因其布置在园林内，与地形、山石、绿化等自然环境相结合，创造出一种既庄重又变化的园林气氛，这与紫禁城内主要供治政之用的宫殿迥然不同。

- **北京颐和园仁寿殿**
  仁寿殿位于东宫门内宫殿区的中部，面阔七间，进深五间，四周带回廊，是颐和园宫殿区体量最大的殿堂。

  Hall of Benevolence and Longevity, the Summer Palace, Beijing

  Entering the East Palace Gate, we will find the Hall of Benevolence and Longevity (*Renshou Dian*) lying in the middle part of the palace area. With a space of seven bays by five bays and winding corridors on the four sides, the Hall of Benevolence and Longevity is the hall with the largest volume in the Summer Palace.

for gardens, only those royal ones have palace-style architectures. Palace-style architectures are usually large and tall, with the main hall in the middle and side halls sitting symmetrically on its two sides. In front of the main hall, there is usually a wide court or plaza, which sets off a solemn atmosphere. And since all these are situated in a garden, and are integrated with the land, the rocks, the plants and other elements in the natural setting, what it comes to is a solemn but varying garden, which is vastly different from the palaces mainly used for political activities in the Forbidden City.

中国古典建筑的屋顶
The Roofs of Classical Chinese Architecture

硬山式
有一条正脊和四条垂脊，只有前后两面坡，屋顶在山墙墙头处与山墙齐平，没有伸出部分，简单朴素，等级最低，广泛应用于住宅建筑。

Hard Gable
For a hard gable style roof, there is a main ridge and four diagonal ridges. It has only two slopes. The roof flushes with the gable and has no protruding parts. Simple and unsophisticated as it is, hard gable roof ranks the lowest in all kinds of roofs and is widely applied to residential constructions.

- 硬山式屋顶
  Hard Gable Style Roof

悬山式
悬山式的屋顶与硬山式十分相似，只是两侧不像硬山式那样与山墙齐平，而是伸出山墙之外，伸出的部分由下面伸出的檩承托。

Suspended Gable
A roof of the suspended gable style is much similar to that of the hard gable style, except that this roof does not flush with the gable but covers it. The protruding part of the roof is supported by the out stretching purlins.

- 悬山式屋顶
  Suspended Gable Style Roof

庑殿式

前后左右四面斜坡，前后坡屋面相交形成一条正脊，两山屋面与前后屋面相交形成四条垂脊。这是中国古代建筑中等级最高的屋顶形式，最尊贵的建筑物才可使用。

Hip

A hip roof has four slopes. The front and the back slopes meet at the main ridge, while the side slopes meet the front and the back ones at the four diagonal ridges. This style of roof is the highest in rank of all ancient Chinese architectures and is reserved for the most respectable constructions.

- 庑殿式屋顶
  Hip Roof

歇山式

最基本、最常见的一种屋顶形式。前后左右四个坡面，同时在左右两端的坡面上还各有一个垂直面，几个面相交出九条屋脊。

Hip and Gable

The hip and gable roof is one of the most common seen roof styles. It has not only the front, the back, the left and the right, altogether four slopes, but also two vertical slopes on the left and the right. These slopes intersects with each other on nine different ridges.

- 歇山式屋顶
  Hip and Gable Roof

卷棚式

屋面双坡，没有明显的正脊，即前后坡相接处不用脊而砌成弧形曲面。因为这种屋顶线条流畅、风格平缓，所以多用于园林建筑。

Round-ridge

A round-ridge roof has only two slopes with no apparent main ridges. It means the two slopes are connected with a curved surface. Because of the smooth lines and gentle style, such a roof is mostly used for garden architectures.

● 卷棚式屋顶
Round-ridge Roof

攒尖式

攒尖式屋顶用于亭、榭、阁和塔等圆形和正多边形的建筑，其特点是屋顶为圆锥或棱锥形，没有正脊，顶部集中于一点。

Pavilion

Pavilion roof is usually used for round shaped architectures like pavilions, pavilions on terrace and pagodas, and regular polygonal architectures. This kind of roof is characterized by a cone or pyramid-shaped roof, absence of the main ridge and the top focusing on one point.

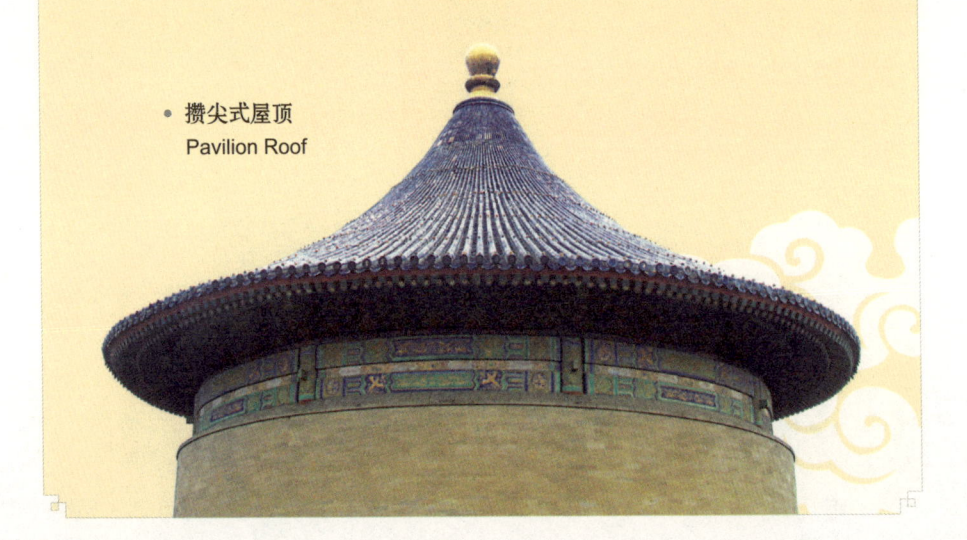

● 攒尖式屋顶
Pavilion Roof

> 厅堂建筑

厅和堂是古代住宅中的主要建筑，也是古典园林中的主体建筑，是园林主人游园和招待宾客的场

> Parlors & Halls

Parlors and halls hold an important position in ancient residences, and they are also the dominating architectures in classical gardens. As the principal scenes of the

- **苏州拙政园远香堂（四面厅）**

   四面厅是园林厅堂中比较高级的一种。屋顶多为歇山式，面阔三间或五间，四面开敞，可以四面观景，不砌墙壁，柱间安装连续的隔扇，檐下设有回廊。四面厅建于园中环境开阔、风景变化丰富的地点，既可在堂上坐观，又可沿廊浏览。苏州拙政园远香堂就属于四面厅，位于中园的正中心，东西南北四个方向都有美景，在室内走动观看，好似一幅中国山水长卷。

   Drifting Fragrance Hall, Humble Administrator's Garden, Suzhou(Four-sided Hall)

   Four-sided halls rank relatively high in garden halls. Their roofs are mostly hip and gable roofs. They are usually three or five bays wide, open on four sides without walls, which allows sightseeing on four sides. There are consecutive partition boards between pillars, and a winding corridor under the eaves. Four-sided halls are built where there is broad environment and various views. The scenery can be enjoyed both in the hall and along the corridor. The Drifting Fragrance Hall of the Humble Administor's Garden in Suzhou is a four-sided hall. It lies in the heart of the garden with beautiful scenes on four sides. The scenes come together as a scroll of Chinese landscape painting if the visitors walk inside the hall to watch.

所，也是园林的主要景观。所以厅堂一般都高大宏敞，正面向南，装修考究，陈设精丽。按厅的功能、构造特点可分为四面厅、鸳鸯厅、花厅、荷花厅等。

garden, parlors and halls are places where the owner enjoys the garden views and entertains his guests. Generally, these parlors and halls are tall buildings facing due south with elegant decoration and fine furnishings. According to their functions and structures, parlors and halls can be grouped into four-sided halls (*simian ting*), mandarin-duck halls (*yuanyang ting*), flower halls (*hua ting*), lotus halls (*hehua ting*), etc.

- 北京涛贝勒府宜安宫（花厅）(图片提供：FOTOE)

园林中的花厅是主人进行宴客、会友、听曲等社交活动的地方，位置多靠近住宅，环境安静，与主要景区隔离，装饰陈设力求精美。厅前自成庭院，内种植数株花木，散点石峰，形成幽静雅致的环境。花厅的屋顶多为卷棚顶形式，在私家园林中比较多见。

Yi'an Palace, *Beile Tao*'s House, Beijing (Flower Hall)

Flower halls in the gardens are lieu for the owner to conduct social activities like serving the guests, meeting with friends and entertaining people with operas. Flower halls are usually located close to the dwelling house in a tranquil environment separated from the main scenery sections. The decoration of the flower halls aims for refinement. In front of the flower hall, there will be a court decorated with flowers, trees and rocks. Tranquility and elegance arise spontaneously. Flower halls are relatively commonly seen in private gardens. Their roofs are mostly round-ridge roofs.

面阔

中国古代木构建筑中，正面相邻两根檐柱之间的水平距离称为"开间"，各开间宽度的总和称为"通面阔"。不同的建筑开间大小有所不同。例如一幢建筑"面阔三间"，就是说其横向有四根檐柱，宽度有三开间。

Building Width (*Miankuo*)

For traditional Chinese wooden architectures, the horizontal distance between two neighboring frontal eave columns is called a Bay (*Kaijian*). The sum of all the bays is called the Total Building Width (*Zong Miankuo*). The bays of different architectures vary in size. A building whose building width is three bays has four eave columns in the frontal side, and thus is three bays wide.

隔扇

中国古代木建筑门的一种，上部有用棂条组成的花格，用于分隔室内外或室内空间。园林中的隔扇通常做得比较轻薄，而且镂空较多，大片的窗格花纹极富变化，增强了房屋的美感和园林的意境。根据建筑物开间的尺寸大小，一般每间可安装四扇、六扇或八扇隔扇。

Partition Board

Partition board is a kind of doors for ancient Chinese buildings used to separate the inside from the outside or different spaces in the inside of the room. It has traceries composed of lattices in its upper part. Partition boards in the gardens are generally lighter and thinner, with more hollow parts. Large traceries have patterns full of variety, which enhance the artistic appearance of the house and the poetic imagery of the garden. Four, six or eight partition boards are installed in one bay according to the size of the bay.

- 北京颐和园仁寿殿的隔扇门
  Partition Boards in the Hall of Benevolence and Longevity in the Summer Palace, Beijing

- 南京瞻园静妙堂（鸳鸯厅）

鸳鸯厅是江南园林厅堂的常见形式之一，是一种较大型的厅堂，一般面阔三间或五间。鸳鸯厅最大的特色就是将整个空间对称地分成前后两部分，中间用隔扇等隔开，似两厅合一，故称"鸳鸯厅"。一般北厅面向园中主景，可通过月台临水看山，夏天乘凉；而南厅阳光充足，适于冬、春两季使用。静妙堂是南方园林中常见的两面临池的鸳鸯厅式建筑，南部的建筑格调清新淡雅，小巧玲珑，而北部的建筑格调粗犷豪放，古色古香。

Hall of Tranquility and Prettiness, *Zhan Yuan* Garden, Nanjing (Mandarin-Duck Style Hall)

Mandarin-duck style halls are also commonly seen in the gardens on the Yangtze Delta. They are relatively larger, with a width of three or five bays. The most distinguished characteristic of the mandarin-duck halls is that the space is symmetrically divided into two parts—the front and the back—with partition boards or other material standing in the middle for separation. The two parts are almost identical, just like the mandarin ducks that always appear in a pair. So this kind of halls is called mandarin-duck halls. Usually the northern part of the hall faces the major scenery spots of the garden; rockeries can be seen from its terrace which is also a place for cool breeze in the summer. The southern part, on the other hand, is sunny, perfect for winters and springs. The Hall of Tranquility and Prettiness (*Jingmiao Tang*) is a mandarin-duck style hall with water on both sides, which is common for southern gardens. Architectures in the southern part are quite elegant and dainty, while those in the northern part are boldly outlined and primitively antique.

- 苏州留园涵碧山房（荷花厅）

荷花厅是江南园林中常见的厅堂之一，形式相对简单，面阔多为三开间，内部多处理成单一空间，南北两面开敞，东西两侧用山墙封闭，或于山墙上开窗取景。荷花厅以观水景为主，多临近水池而建，厅前有宽敞的平台，以便接近水面。池中常常植有荷花，是观赏池水荷花的佳处。此图中左边的二层建筑为明瑟楼，右边则为涵碧山房。涵碧山房为园中部主要建筑，高大宽敞，陈设朴素，厅前平台宽广，依临荷花池，是盛夏纳凉赏荷的好去处。

Hanbi Mountain Villa, Lingering Garden (*Liu Yuan*), Suzhou (Lotus Hall)

Lotus halls are one of the commonly seen parlors and halls in the gardens on the Yangtze Delta. The structure is relatively simple, with building width mostly as three bays, which generally link with each other to form a single space. The north and south sides are open, while the east and the west sides are closed with gables. Sometimes, windows are opened on the gables. Lotus halls are mainly reserved for water views. A lotus hall is usually built beside a pool, with a spacious terrace stretching for the water. Since lotus is planted in the pool, the lotus hall beside is an ideal place to admire lotus. The two storied building on the left of the picture is the Pellucid Tower (*Mingse Lou*) while the architecture on the right is *Hanbi* Mountain Villa (*Hanbi Shanfang*). It is the major architecture in the middle part of the whole garden. The hall is tall and wide, with simple furnishing and a spacious terrace beside the pool. It is an excellent place for breeze and lotus in a hot summer.

> 楼阁建筑

楼和阁，都是园林中高大多层的木结构建筑，造型多样。楼，是指两层以上的房屋；阁，指下部架空、底层高悬并四周开窗的建筑。由于楼与阁形制相似，界限已不严格，因而同一建筑形式有时称为"楼"，有时又称为"阁"。

楼作为主景时，造型突出鲜明；如帝王宫苑中，常以楼作为构图中心，崇楼高阁，多重檐屋，轮廓造型丰富，金碧辉煌，如颐和园佛香阁。楼也可作配景，半隐半现，如南方园林中楼的尺度小巧，面阔三五间，歇山式或硬山式屋顶，朝向园内的一面装栏杆、长窗，山面白粉墙上辟洞门或设砖框漏花窗，虚实相映，轻快活泼。楼梯可设在室内，也可由室外假山云梯直达二层。

> Storied Buildings & Towers

Both *Lou* (storied buildings) and *Ge* (towers) are multiple-storied wooden constructions in gardens. They are rich in forms and styles. *Lou* is a house with two or more stories. *Ge* is built on stilts and has windows on four sides. Since they are similar in appearance, there is no strict boundary between *Lou* and *Ge*. Therefore, the architecture of the same style would sometimes be referred to as *Lou* and other times as *Ge*.

While *Lou* is set as the major scene of the garden, it has a distinct appearance. For example, in royal gardens, *Lou* is usually framed as the center of the picture. They are lofty buildings with multi-eave roof, gorgeous profile and splendid glory. One such building is the Tower of Buddhist Incense (*Foxiang Ge*) in the Summer Palace. *Lou* can also be set as the minor scene in gardens, partly hidden and partly visible. In southern gardens, *Lou* is usually small and

阁的功能与位置同楼相仿，有时依山临水建一层，也称"阁"。阁的造型比楼更为轻盈、通透，四面开窗，平面常呈四方形或对称多边形，如六边形、八边形等，屋顶作歇山式或攒尖式。

exquisite, with a width of three or five bays, and a roof of hip and gable or hard gable style. It has a balustrade or a long window on the side facing the garden center. There are portals or latticed windows on the wall. The distant sceneries and the nearby ones complement each other, creating a cheerful and lively air. Stairs could be placed inside the building or on the rockery beside it.

The function and location of *Ge* are similar to that of *Lou*. A one-story building at the mountain and by the water is sometimes referred to as *Ge*. Compared with *Lou*, *Ge* is a more penetrating and lightsome space with windows on four sides. The cross section of *Ge* is usually a square or a symmetric polygon, such as hexagonal, octagonal, etc.

Ge usually has a hip-and-gable style or pavilion style roof.

- 北京故宫御花园延晖阁
  延晖阁位于御花园西路建筑的最北端，为上下两层，体量高大，装饰典雅，在苍松翠柏的掩映下更显得壮丽辉煌。
  The Belvedere of Prolonging Splendor, the Imperial Garden in the Forbidden City, Beijing
  The Belvedere of Prolonging Splendor (*Yanhui Ge*) lies at the north end of the constructions of the road west to the Imperial Garden. It has two stories and a large volume. Its elegant profile is added with magnificence and splendor in the charming contrast with the green pines and verdant cypresses.

- **苏州拙政园见山楼**

见山楼位于拙政园中部水池的西北角，是一座三面临水的楼阁，上下两层，楼体稳重、朴实。楼的东北角有曲桥与池岸相通，将楼与水、岸联系起来，便于游人登楼观景。

Mountain-in-view Building, Humble Administrator's Garden, Suzhou

The Mountain-in-View Building (*Jianshan Lou*) is located in the northwest corner of the pool in the middle part of the Humble Administrator's Garden. It is a two-story building with three sides facing water, solid and plain. The northeast corner of the building is linked to the shore by a zigzag bridge. Tourists can walk over the water from the shore and then atop the building to marvel at the garden.

> 轩馆斋室建筑

> Xuan, Guan, Zhai and Shi

轩、馆、斋、室是园林中数量最多的建筑物，属于中等大小的建筑，对园林空间的组织、园林景观

Xuan, guan, zhai and shi are the most abundant architectures in gardens. They are medium in size but play an important role in the organization of the garden space and outlooks. Compared with parlors and

- **苏州留园闻木樨香轩（轩）**

  轩是四面空透的建筑，常建于园中次要位置，环境较安静，或是作为观赏性的小建筑。轩的形式非常多样。有种临水而建的敞轩，临水一侧完全开敞，仅在柱间设美人靠，供游人倚坐，与水榭相近，但一般不像榭那样伸入水中。闻木樨香轩位于留园中部的假山上，是秋季赏桂花的佳处。

  Osmanthus Fragrance Pavilion, Lingering Garden, Suzhou (*Xuan*)

  Xuan is open on four sides and usually occupies a less important place where it is quieter. It sometimes serves mainly as a petite architecture for appreciation. *Xuan* has diversified forms. One kind is built beside the water; the side facing the water is fully open, with only the Beauty's Chairs (*Meiren Kao*) between the pillars. Such is the Full-open *Xuan* (*Chang Xuan*). It looks similar to waterside pavilions but it does not stand in the water. The Osmanthus Fragrance Pavilion (*Wenmuxixiang Xuan*) is built on a rockery in the middle part of the Lingering Garden. It is a perfect place to admire osmanthus in autumns.

面貌塑造起着重要的作用。在个体造型、布局方式、建筑与环境的结合上，都表现出比厅堂更多的灵活性。有的布局极为开敞，以建筑个体形式深入到自然环境之中，成为风景画面的重要点缀；有的与厅堂主体建筑组成建筑群组；有的还独立组成一个环境幽静的庭园空间。

halls, studios and pavilions, etc. are more flexible in shapes, layout, and combination with the surrounding environment. Some are with extremely open layout and integrate with the natural environment as an individual architecture, becoming a vital embellishment for the overall landscape; some become a part of the construction complex together with major architectures like parlors and halls; and some stand alone as a reclusive space in tranquility.

- **苏州网师园琴室（室）**

  室，多为园林中的辅助性用房，配置于厅堂的边沿。苏州网师园中的琴室，是一开间的小室，位于一个独立的小院中，庭前砌有湖石壁山，配以丛竹，环境幽静闲适，是弹琴习唱的地方。

  Music Room, Master-of-nets Garden, Suzhou (*Shi*)

  Shi is mostly used as supporting constructions in the gardens, at the side of parlors and halls. The Music Room in the Master-of-Nets Garden is a small building within an independent courtyard. The court is decorated with lakeside rocks and bamboos. Such a leisurely and comfortable environment is a place for playing music and singing.

- **苏州拙政园玲珑馆（馆）**

馆的用途与厅堂相仿，可作游宴听曲、起居会客之用，在园中位置不甚显著，规模有大有小，体型、布置也较为灵活，或面向庭院，观赏山石花木，或临水倚楼，随意经营。江南园林中的馆一般建筑尺度不大，常与居住部分和主要厅堂有一定的联系。玲珑馆位于拙政园中部园区的东南部，三开间，四面隔扇门窗。门窗棂格花纹全部为冰裂纹，与馆前地面上的冰裂纹相互呼应，共同表达了玲珑馆冰清玉洁的精神与高洁品味。

Exquisite Hall, Humble Administrator's Garden, Suzhou (*Guan*)

Guan has similar functions with parlors (*ting*) and halls (*tang*); it can be used for banquets, opera enjoyment, daily life and social activities. Its position in the garden is inconspicuous and its size, form and layout are varied. Some face the courts and embrace the views of rockeries and plants; some overlook the water while lean against other towers. *Guan* in gardens on the Yangtze Delta is usually small in scale, connected with residential parts and major halls or parlors. Exquisite Hall (*Linglong Guan*) locates itself on the northeast part in the central area of the Humble Administrator's Garden. It is three bays wide, with partition boards and windows on four sides. The lattice patterns on both the partition boards and windows are cracked-ice patterns which echo with those on the ground in front of the hall, representing the pure spirit and noble taste.

● 苏州网师园集虚斋内景（斋）

园林中的斋，多为书屋之类的建筑，一般处于静谧、封闭的边落小庭园中，一屋一院，与外界隔离，相对独立，形成统一完整的空间气氛。集虚斋在网师园中心水池的东北部，竹外一枝轩的后面，是一座二层的楼房，上下两层的面阔均为三开间，前部安装木质隔扇。当初，这里楼上是女儿闺阁，下为读书之处，窗外种有丛丛青竹，确有虚静清明之意。集虚斋体量高大，又居于较好的位置，所以登楼观景的视野也非常好。

Interior of Meditation Study, Master-of-nets Garden, Suzhou (*Zhai*)

In the gardens, *Zhai* is usually used as studies and placed in a quiet, reclusive and remote yard. This single house with an exclusive courtyard is separated from other parts of the garden and is relatively independent in its own aura. The Meditation Study (*Jixu Zhai*) lies to the northeast of the central pool in the Master-of-nets Garden, behind the Pavilion of the Bamboo Branch Beyond (*Zhuwai Yizhi Xuan*). This two-storied building is three bays wide on both stories, with wooden partition boards on the front. Upstairs was once the bower for maidens and the downstairs rooms were for reading books. Out of the window, there plant clumps of bamboo, setting off a clear and bright atmosphere. The Meditation Study stands high and occupies a relatively good position, so it offers a perfect view on the top.

临水而建的榭和舫
Waterside Pavilions and Boat-like Houses

- 苏州拙政园芙蓉榭（榭）

榭又称"水阁"，建于池畔，平台挑出水面，以便游人观览园林水景。榭的临水面开敞，设有栏杆。其基部一半在水中，一半在池岸，跨水部分多做成石梁柱结构，较大的水榭还有茶座和水上舞台等。芙蓉榭位于拙政园东部园区池水东岸，是一座临池水而建、四面开敞的水榭，并且建筑体大部分临于水上。水榭屋顶四角飞翘，体态轻盈小巧，倒影入水中，更有一种灵动与飘逸之美。

Waterside Lotus Pavilion, Humble Administrator's Garden, Suzhou (Xie)

Xie is also referred to as *Shui Ge* (waterside pavilion). It is built on the bank of a pool, with its platform stretching out to the water, which allows tourists to take a better view of the water scenery of the garden. Half of it stands over the water and half on the bank. The half over the water is usually supported by stone posts. Bigger *xie* will have a tea lounge or a stage for performances. Waterside Lotus Pavilion (*Furong Xie*) lies on the east bank of the pool in the eastern part of the Humble Administrator's Garden. It is a waterside pavilion the most part of which is over the water. Its four sides are fully open and each of the four angles up flies. The dainty reflection of the pavilion on the water surface adds to its artistry and elegance.

- **北京颐和园清晏舫（舫）**

舫是仿照船的造型建在水面上的建筑物，供游人游玩宴饮、观赏水景之用。舫的前半部多三面临水，船首常设有平桥与岸相连，类似跳板。通常下部船体用石料，上部船舱则多用木构。舫像船而不能动，所以又名"不系舟"。游人身处其中，能取得仿佛置身舟楫的效果。北京颐和园昆明湖西岸边有一只汉白玉石制成的石舫，本是乾隆时的旧物。慈禧重建颐和园时，在原船上加盖了两层西洋式楼阁，增设机轮，舱内墁花砖，镶嵌五色玻璃，陈设西洋桌椅，取名"清晏舫"。

Marble Boat, the Summer Palace, Beijing (*Fang*)

Fang is the imitation of a boat and it is built on the water. It is used for feasts and sightseeing. The front part of a *Fang* is usually surrounded by water on three sides. The cabin on the upper part is mostly made of wood. *Fang* looks like a boat, but it cannot float around on the water; it is fixed, so it is also called Boat without an Anchor (*Buji Zhou*). People on it will feel like that they are in sail. On the west bank of *Kunming* Lake of the Summer Palace, there laid a boat made of white marble which was left from Emperor Qianlong's Reign. When *Cixi* (Empress Dowager) rebuilt the Summer Palace, two stories of western styles building and wheels were added to it. The floor inside the cabin is paved with tiles, and the walls are beset with stained glass. It is furnished with foreign style chairs and tables and given the name *Qingyan Fang*.

> 游赏建筑

亭

亭是园林中用得最多的游赏建筑，可休憩凭眺，遮阳避雨，也是园林风景的重要点缀。亭以造型小巧秀丽、玲珑多姿为特色，选材不拘，布设灵活，可建于山上、林中、路旁、水际，形式、比例因地制宜。亭一般不设门窗，柱间作半墙或平栏，设坐槛、鹅颈椅或栏杆，供休憩。檐枋下悬有挂落，也有柱间作洞门或安门窗的。

> Landscape Constructions

Pavilions

Pavilions are the most frequently used construction in gardens. They not only serve as good places to rest and enjoy the scenery without sunshine or rain, but also the important ornament in the garden. Pavilions are characterized by being small and exquisite. The selection of materials can be free and the design of the pavilions is very flexible. They can be built on the hills, in the woods, by the roadside, or around the lake. Generally, the pavilions are without doors and windows. The half wall or flat balustrade are built between the pillars and there are goose-neck-like chairs or railings for rest. There are hanging fascia hanging from the architrave and sometimes doors or windows can be found between the pillars.

- 北京北海见春亭（圆亭）

建筑平面为圆形、顶部也为圆形攒尖的亭子，叫作"圆亭"，形制活泼，在中国园林中广为应用，有许多不同的式样。其中以单檐圆亭、重檐圆亭为豪华的亭式。见春亭位于北京北海公园琼岛的东坡之上，是一座八柱单檐圆亭，紧贴山石而建，四周树木繁茂，是观赏园内春景的佳处。

Enjoy Spring Pavilion of *Beihai* Park in Beijing (Round Pavilion)

The architectural plane is round and the roof is round pavilion roof, called round pavilion (*yuan ting*), which is very lovely and is extensively applied in Chinese gardens. There are many styles of this kind of pavilion, among which the single eave roof and double eave roof are splendid styles. Enjoy Spring Pavilion (*Jianchun Ting*) is situated on the eastern slope of *Qiongdao* in *Beihai* park in Beijing. It is an eight-pillared single eave round pavilion, built along the hills and stones, with flourished trees around. And it is really a perfect place for enjoying spring sceneries.

- 绍兴兰亭"鹅池"碑亭（三角亭）

三角亭只用三根支柱，在结构上显得最为轻巧。绍兴兰亭"鹅池"三角形碑亭为石质，亭中立着刻有"鹅池"二字的石碑，传说"鹅池"两字为大书法家王羲之、王献之父子的手笔。

Goose Pond Triangle Stele Pavilion in *Lanting* of Shaoxing (Triangle Pavilion)

The triangle pavilion is supported with only three pillars, which makes it the most simple and handy pavilion. Goose Pond Triangle Stele Pavilion (*E'chi SanjiaoxingBei Ting*) in *Lanting* of Shaoxing is made of stone. There are two characters (goose pond) on the stone stele. It is said that "goose pond" are the handwriting of two famous Chinese calligraphers, Wang Xizhi and his son Wang Xianzhi.

- 苏州拙政园梧竹幽居亭（方亭）

方亭是建筑平面为方形的亭子，有正方形、长方形两大类，其中正方形的亭子较普遍。方亭的顶子多为方形攒尖顶，也有采用歇山顶、悬山顶、硬山顶、十字顶式的。方形形态端庄，结构简易，可独立设置，也可与走廊结合为一个整体。梧竹幽居亭位于拙政园中部景区的东北角，平面为方形，四角攒尖顶，亭子四面各有一个圆洞门，聚集了亭外四面的景色。

Tranquil Dwelling with Phoenix Trees and Bamboo Pavilion in the Humble Administrator's Garden in Suzhou (Square Pavilion)

With the architectural plane as tetragon, there usually are two types, square and rectangle, with square ones more popular. The roof of square pavilions is mostly square pyramidal roof, but gable and hip roof, overhanging gable roof, flush gable roof and cross gable roof are also used sometimes. Square shape is very dignified and the structure is simple. The square pavilions can be designed independently or emerge with the corridors as a whole. Tranquil Dwelling with Phoenix Trees and Bamboo Pavilion (*Wuzhu Youju Ting*) is located at the northeast corner in the middle scenic zone of the Humble Administrator's Garden. The architectural plane is square and the roof is square pyramidal roof. On each side of the four sides, there is a round door which gathers the all-round views.

- 苏州拙政园荷风四面亭（多角亭）

多角亭是建筑平面为多角形的亭子，主要有五角亭、六角亭、八角亭等，其中六角亭和八角亭较为常见。建于拙政园中部园区水池之上的荷风四面亭是一座单檐六角攒尖顶小亭，亭檐高翘，开敞通透，体态轻盈。亭四面环水，是中部园区的中心景观，也是南北景区重要的连接点。

Lotus Wind from All Sides Pavilion in the Humble Administrator's Garden In Suzhou (Polygonal Pavilion)

Polygonal pavilion means the architectural plane is polygonal. Usually, there are pentagonal, hexagonal and octagonal pavilions, with hexagonal and octagonal ones more common. Built around the pond in the middle scenic zone of the Humble Administrator's Garden, Lotus Wind from All Sides Pavilion (*Hefeng Simian Ting*) is a single eave pavilion with hexagonal pyramidal roof. The eave of the pavilion is light, open and graceful. The pavilion is surrounded by water. And it is the central scenic spot in the middle scenic zone and the important connection point of the south and north scenic zones.

● 北京北海枕峦亭（多角亭）

枕峦亭位于北海静心斋内，建于一座假山顶端。此亭小巧别致，为单檐八角攒尖顶，远看宛如一朵盛开的莲花。亭上匾额"枕峦亭"为清代光绪帝御笔。

Sleeping on the Hill Pavilion in *Beihai* Park of Beijing (Polygonal Pavilion)

Sleeping on the Hill Pavilion (*Zhenluan Ting*) is located in the Tranquil Heart Studio (*Jingxin Zhai*) in *Beihai* Park and is built on the top of a rockwork. This pavilion is small but exquisite and novel. It is single eave with octagonal pyramidal roof, just like a full-blown lotus seen from afar. The characters Sleeping on the Hill Pavilion (*Zhenluan Ting*) is the handwriting of Emperor Guangxu in the Qing Dynasty.

● 苏州狮子林扇亭

平面呈扇面形的亭称为"扇亭"，造型优美独特，在园林中也常采用。苏州狮子林西南园墙角的扇亭，建于曲尺形的两廊之间，与廊贯通，亭后空间辟为小院，布置竹石，犹如一幅小图画，显得十分雅致。

Fan-Shaped Pavilion in Lion Forest Garden in Suzhou

The pavilion with a sectorial architectural plane is called Fan-shaped Pavilion (*Shan Ting*). The profile is elegant and unique, and is often adopted in the gardens. The Fan-shaped Pavilion in the south-west corner of Lion Forest Garden was built between two L-shaped corridors. The pavilion is connected with the corridors and the space behind it is set as a yard with bamboo and stones, just like a picture, looking very elegant.

- 苏州怡园四时潇洒亭（半亭）

半亭，即紧靠墙廊只筑半个亭，其余半个亭则化入墙廊之中。四时潇洒亭就是一座著名的半亭，依曲廊而建，亭内一面设粉墙，墙上开圆形洞门，设计十分精巧而脱俗。

Four Seasons Fine Pavilion in the Garden of Pleasant in Suzhou (Half Pavilion)

Half pavilion means there is only half pavilion connecting closely to the corridor wall, while the other half is fading in the wall. Four Seasons Fine Pavilion is a very famous half pavilion. It is built along the zigzag veranda. One side of the pavilion is painted pink and there is one round gate on the wall. The design of it is refined and ingenious.

- 北京天坛鸳鸯亭（套方亭）

天坛的鸳鸯亭属于套方亭，又称"方胜亭"，由两个方亭沿对角线方向组合在一起。套方亭一般组合方式是在正方亭相邻的两个边上各取中点，以连接这两点的斜线作为两个正方亭的公用边。

Mandarin Ducks Pavilion (*Yuanyang Ting*) in the Temple of Heaven in Beijing (Interlocking Diamond-shaped Lozenges Style Pavilion)

Mandarin Ducks Pavilion in the Temple of Heaven is an interlocking diamond-shaped lozenges style pavilion, which is also called *Fangsheng Ting*. It is a combined pavilion of two square pavilions along the diagonal line. The usual way of combining the two square pavilions is finding the mid points of the two adjacent edges, and taking the slanting line connecting the two mid points as the common cut for the two square pavilions.

- **苏州网师园月到风来亭（水亭）**

  水亭一般尽量贴近水面修建，突出于水中，三面或四面为水面所环绕，同时在亭边水中散置叠石，以增添自然情趣。月到风来亭位于网师园内彩霞池西，三面环水，一面与曲廊相连。亭为六角攒尖顶，檐角飞翘，纤巧美观。

  The Moon Comes with Breeze Pavilion in the Master-of-nets Garden in Suzhou (Water Pavilion)

  Building a pavilion along the water often requires the pavilion being built near the water surface, prominent among the water and three sides or four sides surrounded by the water. At the same time, rocks are cluttered around so as to add the natural feeling. The Moon Comes with Breeze Pavilion (*Yuedaofenglai Ting*) is situated in the west of Rosy Clouds Pond in the Master-of-nets Garden. Three sides of the pavilion are surrounded by water while the fourth side is connected with the zigzag veranda. It is a pavilion with hexagonal pyramidal roof and upwarped eaves and is very delicate and beautiful.

- **苏州留园濠濮亭（水亭）**

  濠濮亭建于留园中部水池的东岸，几乎全部凌于水上，三面临水，一面连着石岸，亭下各角以乱石为柱支撑亭身。小亭空间较开敞，便于观赏四面景致，亦可俯视水中游鱼。

  ***Haopu* Pavilion in the Lingering Garden in Suzhou (Water Pavilion)**

  Haopu Pavilion was built in the east shore of the middle part pond in the Lingering Garden. The whole pavilion is almost on the water, with three sides surrounded by water, the last one side is connected to the rock shore. There are rocks under each end of the pillar to support the pavilion. The pavilion is quite spacious and is convenient for appreciating the all-round views and looking at the fish in the pond.

● 苏州拙政园待霜亭（山亭）

山亭一般选择宜于鸟瞰远眺的地形，且眺览范围越空阔越好，山巅、山脊、山腰上是特别宜于建亭之处。待霜亭在拙政园中部水池东面的小岛上，是一座六角形单檐小亭，因小岛山势高耸，故能远眺。待霜亭四周原来植有橘树数十株，景致自然，具有田园风味。

The Orange Pavilion in the Humble Administrator's Garden in Suzhou (Hill Pavilion)

Building a pavilion on the hill usually requires the landform providing a bird's-eye view. And more spacious the view range is, the better it is for building a pavilion. The top, the ridge of the hill or the hillside are often the appropriate places for building pavilions. The Orange Pavilion is on the islet on the east shore of the middle part pond in the Humble Administrator's Garden. It is a small pavilion of single eave and hexagonal pyramidal roof. Because the the hill on the islet stands tall, people can look afar into distance from the pavilion. There used to be about ten orange trees around The Orange Pavilion. The views are natural and pastoral.

● 苏州拙政园天泉亭（井亭）

天泉亭位于拙政园东区偏东北处，亭子的上层檐略小，下层檐稍大，整体造型稳重，但檐角飞翘，具有灵巧秀美之气。天泉亭中有一口井，井水终年不竭，甘甜清洌。

Heavenly Spring Pavilion in the Humble Administrator's Garden in Suzhou (Well Pavilion)

Heavenly Spring Pavilion is located in the northeast part of the east scenic zone of the Humble Administrator's Garden. Compared to the lower eave, the upper eave is small, which makes the overall structure steady. But the eaves are upwarped, so at the same time, it is delicate and elegant. There is a well in the Heavenly Spring Pavilion. The water in the well is inexhaustible and very sweet.

中国园林中的建筑 Architectures in Chinese Gardens

- 北京太庙井亭

井亭因亭中有一口井而得名，一般是因井水特佳，才会盖亭。井亭多见于宫廷中和皇家园林内，民间园林中的井亭较为少见。

Well Pavilion in Beijing Imperial Ancestral Temple (*Taimiao*)

Well Pavilion got its name because of the well in the pavilion. Usually, the pavilion is built only when the well water is very good. Well Pavilion can be commonly seen in imperial or palace gardens, not usually be found in the folk gardens.

- 南京燕子矶御碑亭

碑亭是为保护石碑而建的亭子，多见于景观园林之中。园林内的碑亭并不像陵墓碑亭那样封闭，多采用四面开敞形式，尽显园林建筑的轻巧之美。御碑亭建于清乾隆三十年（1765年），为双层六角重檐结构，重檐翘角系有风铃，风吹铃响，清音悦耳，为燕子矶一景。

Swallow Rock Imperial Stele Pavilion in Nanjing

Stele pavilion is built for the protection of stele, which can be often seen in landscape gardens. The stele pavilion is not that closed as mausoleum tablet. It usually employs the all-round open way of design, which can illustrate the special elegant beauty of garden architecture. The Imperial Stele Pavilion (*Yanziji Yubei Ting*) was built in 1765 with the structure of double eave hexagonal pyramidal roof. There are wind-bells on each end of the double eave. The sound of the wind-bell is very pleasing to the ear, which makes it a special view in Swallow Rock.

● 扬州瘦西湖五亭桥（桥亭）

桥上建亭，是我国古典园林建筑处理常用手法之一。桥亭结构各异，既要与桥身相和谐，又要与全园的建筑风格相统一。站在桥亭上，可以更好地观赏水景，又能防止雨淋日晒，桥亭可以设置坐凳栏杆，方便游人休息。瘦西湖是江苏扬州著名的景观园林，园中景观以五亭桥最为闻名。亭在桥上，桥下有5个拱形桥洞，洞洞相连。这座五亭桥是仿北京北海北岸的五龙亭而建，设计更加玲珑精巧，更具江南园林的特色。

Five-pavilion Bridge in Slim West Lake in Yangzhou (Bridge Pavilion)

Building pavilions on bridge is one of the commonly used techniques in Chinese classic garden architecture. The structures of the bridge and the pavilion are different, which requires the pavilion not only be harmonious with the bridge, but also fit in the overall architectural style of the garden. Standing on the bridge pavilion can bring people a better view of the waterscape. What's more, people can be protected from burning sunshine or rains and can rest on the chairs along the railings. Slim West Lake (*Shouxihu*) is a very famous landscape garden in Yangzhou, Jiangsu Province; and Five-pavilion Bridge (*Wuting Qiao*) is the most famous scenic spot in it. The pavilion is on the bridge, while under the bridge, there are five connected archways. The Five-pavilion Bridge imitates the Five-dragon Pavilion (*Wulong Ting*) in *Beihai Park of Beijing*, but with more delicate design of gardens on the Yangtze Delta.

沁秋亭地面上的水槽

流杯亭，因亭内设有石制的流杯渠而得名。沁秋亭是中国古代园林中现存的流杯亭之一，是恭王府主人宴请宾客的地方，每逢春秋之际，或仲春花月夜，园主人便邀请客人在此饮酒赋诗。

Water Channel on the Floor of the *Qinqiu* Pavilion

Floating cups pavilion (*Liubei Ting*) is named so because of the stone-made floating cups channel. The *Qinqiu* Pavilion is one of the floating cups pavilions from ancient Chinese gardens. It was the place where the masters of the Prince Gong's Mansion used to treat guests. On the spring or autumn days, the princes invited the guests to drink and compose poems here.

北京恭王府花园沁秋亭（流杯亭）
Qinqiu Pavilion in the Prince Gong's Mansion in Beijing (Floating Cups Pavilion)

曲水流觞

曲水流觞，最初是中国古人每年农历三月三日在水边进行的祈福活动，后来逐渐发展成为文人临水赋诗、饮酒赏景的风雅之举。"觞"是古代的一种饮酒器，类似今天的酒杯，通常为木制，小而体轻，底部有托，可浮于水中。大家坐在水渠两旁，在上流放置觞，任其顺流而下，停在谁的面前，谁就得即兴赋诗并饮酒。后来，这个活动逐步由室外缩小到在凿有弯曲回绕水槽的亭子内进行。

Qushui Liushang

Originally, *qushui liushang* refers to ancient Chinese's benediction activity along the waterside on the third day of the third lunar month. Gradually, it has become the graceful activity of Chinese literati drinking, versifying while appreciating the view along the waterside. *Shang*, is a kind of drinking vessel in ancient China. It is similar to today's drinking cup, but usually made of wood, and it is small and light. The tray at the bottom of *shang* makes it possible to float on the water. People sit along the two sides of the water channel and put *shang* on the upper reaches, so that it would flow down along the stream. And when it stops in front you, you must drink the wine and versify spontaneously. Afterwards, *qushui liushang* is gradually confined to the pavilions with winding water channels, rather than in the open air.

- 北京天坛双环万寿亭

双环亭是将两个单体圆亭结合在一起形成的组合亭，一般由两个八柱圆亭组成，也可用两个六柱圆亭套在一起组成双环亭。位于北京天坛公园西北隅的双环万寿亭，是一座由两个重檐八柱圆亭组合而成的双环亭，原建于西苑（现中南海），是乾隆皇帝为其母祝寿时所建，1977年移建至天坛公园。

Double-round Longevity Pavilion in the Temple of Heaven in Beijing

Double-round pavilion is a combined pavilion of two round pavilions, which is often composed of two eight-pillared round pavilions, sometimes two six-pillared round pavilions. Located in the northwest part of the Temple of Heaven, Double-round Longevity Pavilion has two eight-pillared round pavilions with double eaves roof. It used to be located in *Xiyuan* (now *Zhongnanhai*), built by Emperor Qianlong of the Qing Dynasty for his mother's birthday. And in 1977, the pavilion was moved to the Temple of Heaven.

廊

廊是走廊，原本是位于厅堂四周的附属建筑，后来成为独立的园林建筑形式之一。廊的造型轻巧玲珑，立面多开敞，也有作漏花墙的。廊柱间砌矮墙，覆砖板，上悬挂落，呈连续装饰，天花常做成各种轩式，整齐美观。

廊的形式有直廊、曲廊、波形廊、复廊，按其位置则有沿墙廊、空廊、回廊、楼廊、爬山廊、涉水

Corridor

Corridor (*lang*) is passage. It used to be the attached construction to the halls, but later it has developed into one of the independent architectural forms in the gardens. The profile of corridors is light, exquisite and open. And sometimes corridors can be used as screen walls. There are seat walls between the pillars and hanging fascia from the roof of the corridors. The roofs have various styles and are very neat and beautiful.

- **上海豫园内的直廊**
  直廊，是走势比较平直的廊子。由于直形的走廊缺少变化，太长会显得单调，所以在园林中大多较为短小。
  The Straight Corridor in Shanghai *Yuyuan* Garden
  The straight corridor, just as its name implies, refers to the unbent passages. Since the lack of change, most of the straight corridors are short, so as to avoid the problem of plainness.

廊等。廊可随形而弯，依势而曲，可蜿蜒山坡，凌空水上，或穿花丛，或入竹林。

- **苏州留园曲廊**

曲廊是园林中最为常见也最富变化的一种廊子。它可以在园中曲折逶迤，自由穿梭，将园林分成各不相同的区域，丰富了园林的景致。留园布局紧凑，各景区用曲廊相连。全园曲廊长达700多米，随形而变，顺势而曲，使园景显得深远而又富于变化。

The Zigzag Corridor in the Lingering Garden in Suzhou

The zigzag corridors are the most popular ones and they are full of changes. They can wind freely and divide the garden into different zones which enrich the views of the gardens. The Lingering Garden is well distributed and the different zones are connected by the winding corridors. The total length of the winding corridors is 700 meters. They wind freely along the landform, which makes the whole garden far-reaching and rich in changes.

Corridors have many forms: straight corridor, winding corridor, waveform corridor and double corridor. According to the position, corridors can be divided into along-the-wall corridor, hollow cloister, two-storied corridor, sloping corridor, and along-the-water corridor. The construction of corridors is flexible. It can wind along the landform and hillside, over the water surface or walk through the flower field and bamboo grove.

• 苏州怡园复廊（图片提供：黄滢/FOTOE）

复廊一般是隔着院墙修建两条长廊，两条长廊都可以行走，因院墙上设有漏窗，人在廊中行走时可看到复廊两侧的景物，增加了园林观赏的层次感。苏州怡园的复廊将园分为东西两大部分，廊的两边都有景物可赏，而两边景物的特征又各不相同。东园以建筑为主，西园以水景为主，从复廊的花窗中看东西两面园景，显得特别幽深曲折。

The Double Corridors in the Pleasant Garden in Suzhou

The double corridors mean there are corridors on each side of the courtyard wall. Because of the hollowed screen wall, people can enjoy the sceneries on both sides while walking on the corridors, which definitely adds the sense of layering of the garden. The double corridors in the Pleasant Garden divide the garden into east and west parts. On both sides, there are very different sceneries for appreciating. The east garden is famous for the architecture while the west garden is mainly of water scenery. While enjoying the sceneries of both sides through the hollowed screen wall, the whole garden is so serene and winding.

• 上海豫园水廊

水廊是临水或跨越水面而建，可分割水面，使景物若隐若现、空间半隔半透，增加水景的变化和层次，使人感到水面开阔，水流不尽。

Water Corridor in *Yuyuan* Garden in Shanghai

Water corridor is built along the water or over the water. It can divide the water surface and make the sceneries partly hidden and partly visible, which adds the changes and sense of layering of the water scenery. People then will have a feeling of open water surface and unlimited water stream.

- 北京北海琼岛爬山廊

爬山廊依山坡的起伏而建，廊内修有台阶或者坡路；廊的顶部可做倾斜式，也可以做成跌落式。爬山廊具有随山势高低起伏的形态，丰富了园林景色。

Sloping Corridor in *Qionghua* Islet in *Beihai* Park in Beijing

The sloping corridor is built along the wave of the hillside. There are steps and sloping pathway within the corridor. The roof of the sloping corridor can be sloping style or falling style. The profile of the sloping corridor similar to the ups and downs of the hill makes the view of the garden richer.

- 扬州何园蝶厅楼廊

楼廊又叫"双层廊""阁道"，是有上、下两层的走廊。楼廊一般多与楼阁相连，组成特别的园林建筑景观。因其较高，视野更为广阔，可以欣赏到更多的园中景致。也有通过楼廊将假山与楼厅相连的。扬州何园的蝶厅楼廊，不仅平面上有单廊、复廊之分，而且立面上也分上下两层。廊总长1500余米。

Butterfly Hall Two-storied Corridor in *Heyuan* Garden in Yangzhou

Loulang is also called two-storied corridor or flying corridor, which has the upper and lower corridors. Usually, the two-storied corridor is connected with pavilion, which creates a special view in garden architecture. The higher position of two-storied corridor not only makes the views wider but also makes it possible to enjoy more scenery. Sometimes, the corridor connects the rockwork and the pavilion. The corridor in *Heyuan* Garden, on the horizontal line, has single and double corridors; while on the vertical line, it has two stories. The total length is more than 1,500 meters.

- 苏州环秀山庄西廊（单面廊）

单面廊是一侧通透、另一侧为墙或建筑所封闭的廊子，又叫"半廊"。通透的一面多面向园林内部，便于观赏园林景致；而另一面的墙可以完全封闭，也可设计为半封闭式，如设置花格或漏窗。环秀山庄问泉亭水池的西边有一座两层的长楼，称为边楼或西楼。楼体的下层前部带有一条非常漂亮的长廊，属于单面廊。廊的前檐下有立柱、设木质矮栏，廊后檐墙面上设有漏窗。墙上的漏窗有方形、圆形、矩形、扇形、横八字形、灯笼形等，窗洞内的形状也各有各的形式，丰富多变。

West Corridor in the Mountain Villa with Embracing Beauty in Suzhou(Single Side Corridor)

Single side corridor, also called half corridor, is the corridor with one side open, while the other side is wall or closed construction. The open side is usually faced with the inner part of the garden so as to be convenient for appreciating the scenery. The other side can be completely closed, or half closed with hollowed screen wall. On the west shore of pond round *Wenquan* Pavilion in the Mountain Villa with Embracing Beauty, there is a long two-storied corridor, called *bianlou* or west corridor. At the forepart of the first story of the building, there is a very beautiful single side corridor. There are pillars to support the corridors in the forepart and you can find wooden fence along the corridor. The closed side has hollowed screen wall with all kinds of shapes: square, round, rectangle, fan-shaped, horizontal eight-shaped and lantern-shaped. Besides, the check lattices are also of many forms, very colorful and varied.

桥

桥在中国园林中是不可缺少的构筑物，亦为园中一景。桥多建于水面较狭窄处，偏于一隅，以保持大片水面空间的完整。桥作近景或中景，使园景更显深邃。园林内常用的桥有平桥、曲桥、拱桥、亭桥、廊桥等类型。

Bridges

Bridges are the indispensable construction and also special sceneries in Chinese gardens. They are mostly built in the less prominent place where the water surface is narrow, so as to sustain the overall completeness of the water space. Being the close range or the medium range scenery, bridges make the garden more artistic. Level bridge, zigzag bridge, arch bridge, pavilion bridge and corridor bridge are often adopted in the gardens.

• 北京颐和园知鱼桥（平桥）

平桥又称"梁桥"，一般是在水中立石柱、石梁、上架石板而筑成，多见于尺度较小的江南园林中。石板桥多低平，贴水面而过，既便于观赏水中的游鱼、莲荷，又令人感到水面比实际宽广。桥上要有木栏或石栏，较为低矮，简洁轻快。知鱼桥在颐和园的谐趣园中，为七孔平桥形式，桥身贴近水面，让游人可以近距离观赏水中游鱼。桥头还立有一座简单的石牌楼，上刻"知鱼桥"字样。

Knowing the Fish Bridge in the Summer Palace in Beijing (Level Bridge)

Level bridge is also called the beam bridge. It often has a flagstone with the stone pillar or stone beam in the water to support and mostly can be found in the comparatively small south gardens. Stone slab bridge is low and flat, just over the water surface, which is not only proper for enjoying the fish and lotus in the water, but also makes the water surface more vast. There are wooden fences or stone fences on the bridge, low but light and neat. Knowing the Fish Bridge (*Zhiyu Qiao*) is in the Harmonious Interests Garden (*Xiequ Yuan*) in the Summer Palace. The bridge is a seven-span level bridge, which is just over the water surface, and visitors can enjoy the fish from a close range. A simple stele with the characters Knowing the Fish Bridge on it is at the end of the bridge.

• 台湾南园曲桥

曲桥实际上也属于平桥，只是桥身曲折，一般有一折、二折至三折、四折乃至九折。曲桥大多架设在园林池水之上，以分隔水面，避免显得单调。南园位于台湾新竹的新埔镇，是一座幽雅的山中园林。园中的木曲桥贴近水面，令人有凌波而行的错觉。

The Zigzag Bridge in the South Garden in Taiwan

The zigzag bridge actually belongs to the level bridge, only more winding. Usually, the zigzag bridge may have one turn, two turns, three or four turns, and even nine turns. It is often built over the pond to divide the water surface, so as to avoid monotony. The South Garden, located in the Xinpu town in Xinzhu, Taiwan Province, is a very serene and elegant garden in the hill. The wooden zigzag bridge in the garden is very close to the water surface, which gives the visitors an illusion of walking on water.

- 北京颐和园玉带桥（拱桥）

拱桥因桥身弯曲如虹，也叫"虹桥"，有单孔、三孔及多孔之分，曲线优美，水中倒影成趣。拱桥桥身高、跨度大，桥下孔洞可行船，多见于较开阔的水面上。也有用山石叠置呈拱状而为拱桥的，并植藤萝缠绕，形若水谷崖洞，别具自然情趣。

玉带桥是颐和园昆明湖西堤六桥之一，是一座单拱的石拱桥，拱券呈抛物线形，桥面高耸，桥下可以通过一般小船。桥身用汉白玉砌成，清隽洁白，桥身高瘦，形似长虹卧波。

Jade Belt Bridge in the Summer Palace in Beijing (Arch Bridge)

Arch bridge is also called rainbow bridge because the profile of the bridge looks like a rainbow. It can be classified into one aperture, tri-aperture and multi-aperture bridges, all with elegant curves and the inverted image in water makes the whole picture more charming. The arch bridge is high with a wide span, and boat could pass through the aperture. This kind of arch bridge can often be seen on the wide and open water surface. But there is another type of arch bridge, which is formed with stack-up rocks into arch shape and Chinese wisteria are twining around it. The whole scene is like a grotto with water, full of natural taste.

    Jade Belt Bridge is one of the six bridges along the west shore of *Kunming* Lake in the Summer Palace. It is a single-aperture arch bridge and the bridge is high; usually a small boat can pass under the bridge. This bridge is made of white marble, so it looks very clean and purely white. The Jade Belt Bridge is high, just like a rainbow lying in the lake.

- 苏州寒山寺外的枫桥

枫桥位于苏州西北的小镇枫桥镇，横跨于运河支流之上。枫桥是一座江南风格的月牙形单孔石拱桥，长39.6米，高7米，宽4.2米，跨径10米，造型古朴优美。

***Fengqiao* Bridge outside *Hanshan* Temple in Suzhou**

Fengqiao Bridge is located in Fengqiao town, a small town in the northwest of Suzhou. It stretches over the branch of the canal. *Fengqiao* Bridge is a south-style crescent-shape single-aperture arch bridge. The length of the bridge is 39.6 meters, the height is 7 meters, the width is 4.2 meters, and the span is 10 meters. The profile of the bridge is of primitive simplicity and very elegant.

> 园林建筑装饰

中国园林建筑的装饰是全方位的，从房顶到地面，从墙壁到门窗，处处都经过设计者精心的安排和设计，使之成为园林艺术的重要组成部分。

景墙

景墙是具有防护功能的围墙、院墙、廊墙等的总称。园墙经过艺

> Decorations of the Architectures

Decorations for the Chinese garden architectures are all-directional — from the ceiling to the floor, and from the walls to the windows. Each detail is carefully conceived and designed and this is what makes decorations a vital part of the garden art.

Landscape Walls

Landscape wall is a general name applied to enclosure walls, courtyard walls, partition walls and other walls for protection but

- 云墙

云墙是中国园林中常用墙体形式之一，又名"波形墙"，即墙头做成起伏波浪形状，线条流畅轻快，富于韵律。墙面常抹白灰，墙头覆小青瓦。

Cloud Walls

Cloud wall (yun qiang) is one of the most frequently used styles of walls. It is also called wave-shaped wall which means it waves on the top. It has sleek and rhythmical lines. The surface is usually whitewashed and the top covered with small green tiles.

术处理后，本身也成为很好的景致，故古代又叫"景墙"。

景墙有云墙、阶梯形墙、漏明墙、平墙等形式。从材料上分，景墙有砖墙、粉墙、石墙、版筑墙等类型，色彩多为白、灰、黄，宜与园林环境谐调。景墙的墙面上常挖

- **承德避暑山庄的虎皮石墙**
  虎皮石墙以不规则的山石块砌成墙，或砌墙的基部石块间以白灰膏勾缝，形成自然的花纹，又显出石材的天然肌理效果，整个墙体具有田园风格。

Masonry Wall with Polygonal Stone, Chengde Mountain Resort
Masonry wall with polygonal stone (Tiger-Skin Walls) is built with irregular stones and pointed with white lime. A natural pattern is thus formed and the original texture of stones is preserved. There is a country flavor in this kind of walls.

with aesthetic treatment. The walls become beautiful features themselves and are thus also called feature walls (*jing qiang*) in ancient times.

Landscape walls vary in forms as cloud walls (*yun qiang*), ladder-shaped walls (*jieti qiang*), lattice walls (*louming qiang*), and flat walls (*ping qiang*), etc. The making of these walls also varies: some are made of bricks; some are made of stones; some are whitewashed; some are built by stamping earth between board frames. Their colors are mainly white, gray, and yellow, which are in concert with the overall garden environment. There are usually latticed windows, empty windows

有漏窗、门洞、空窗，形成明暗、光影、虚实的对比变化，使景墙更具装饰性。有的景墙上有攀缘植物，也很有特色。

and portals on the wall, creating a contrast between the bright and the dark, the light and the shadow, and the real and the virtual. Thus the walls better serve the purpose of decoration. Some landscape walls are woven with climbing plants, which is also distinctive.

- **北京北海的阶梯墙**
  阶梯墙又名"马头墙"，常建于坡地，墙头呈阶梯状错落，轮廓变化，高低相错，有很好的装饰性，可丰富园林风景构图。
  Ladder-shaped Wall, *Beihai* **Park, Beijing**
  Ladder-shaped walls are also referred to as horse head walls. They are usually built on sloping fields. The top of the walls descent like a ladder, offering a varying outline and is very decorative. Such walls can greatly enrich the landscape in gardens.

洞门

洞门，就是在墙体上开设的不装门扇的门洞。洞门的形式多种多样，主要有圆形、方形、横方、直长、圭形、长六角、正八角、长八角、方胜、海棠、桃、葫芦、秋叶、瓶形等。洞门除供出入外，主要用来引导游览路线，还可作为取景画框，以作借景、框景。门后着意布置花木石峰诸景物，犹如画框中的图画小品，自然生动。不管是

Portals

Portals are doors without door leaves on the walls. Portals have a variety of shapes — round, square, rectangle, jade-tablet, long-hexagon, regular octagon, long octagon, interlocking diamond-shaped lozenges, crabapple, peach, calabash, autumn leave, vase, etc. Besides offering a pass, the portals are mainly used to guide the touring route, and are sometimes used as the frame for borrowed sceneries and enframed sceneries. The plants, rockeries, and other landscapes behind the portal

哪种形状的洞门，两侧的墙体都很厚，所以洞门的直径较大，呈圆筒状，使人产生一种幽深之感。

are vivid and natural as a picture or an ornament inside a frame. The walls on both sides are thick for portals of all shapes. The width of the portal is relatively large, setting off a deep and serene feeling.

- **杭州西湖我心相印亭月洞门**
  在墙体上开一圆形洞门，叫"月洞门"或"月亮门"。隔门观景，虚实相间，在不同光影的照射下，景物产生丰富的变化。月亮形门洞反射在景物上的阴影使园林景观更是别有一番情韵。

  Moon-shaped Portal, Pavilion of Heart-linking-to-seart in the West Lake, Hangzhou

  A circular opening on the wall is called a moon-shaped portal or a moon door. When you watch through the portal, the sceneries near and far come into your eyes at the same time. With varying light exposure, the scenes take on rich changes. The shadow created by the moon-shaped portal on other objects lends the garden view a special charm.

- **苏州同里退思园瓶形洞门**
  瓶形门上部较小、下部略大，呈弧形。这种形状的门洞不影响游人通行，还突显出瓶形的门框形式。园林中经常在两处相对的景点上开设瓶形洞门，相互形成对景，别具雅趣。

  Vase-Shaped Portal, Retreat and Reflection Garden in Tongli, Suzhou

  A vase-shaped portal is smaller in the upper part and bigger in the lower part. Such design highlights the shape of a vase without obstructing the way. Two vase-shaped portals are usually employed at the same time on two opposing garden scenes. This opposite scenery carries a refined elegance with it.

- 海棠形洞门

在墙面上开出一个状如四瓣海棠花样的洞门，过往的人们正好从花心处穿行，十分有趣。

Crabapple-shaped Portal

The opening on the door is shaped like a four-petal crabapple flower. People walking through the heart of the flower create an interesting scene.

- 苏州沧浪亭葫芦形洞门

Calabash-shaped Portal of Great Wave Pavilion in Suzhou

漏窗与空窗

漏窗又叫"花窗"，通常是用砖瓦磨制镶嵌在墙面上，构成玲珑剔透的花纹图案，用以装饰墙面。漏窗的形状非常丰富，主要有方形、圆形、六角形、八角形、扇形、菱形、花形、叶形等，窗内纹样也有许多变化。根据所用材料，漏窗还可分为石漏窗、砖漏窗和木雕漏窗。

Latticed Windows & Empty Windows

Latticed windows (*lou chuang*) are also known as flower windows. They are often made of bricks and tiles and then beset onto the walls to form dainty and elegant patterns for decoration. The latticed windows are rich in shapes — square, circle, hexagon, octagon, fan, lozenge, flower, and leaf and so on — and the patterns on the windows vary as well. Latticed windows, while grouped according to the material, include

空窗又名"月洞""窗洞",是在墙面上开挖出不装窗扇的窗孔。空窗有采光通风的实际用途,但更强调点缀园景和组织风景画面的功能。空窗的细部做法、构造与门洞相同,因形状不受限制,所以空窗比洞门更加丰富多样。园林造景中的框景、借景、漏景大多是用空窗来实现的。

stone latticed windows, brick latticed windows and wooden latticed windows.

Empty windows are also referred to as moon openings (*yue dong*) or window openings (*chuang dong*). They are windows without sashes in the walls. Empty windows serve the practical purpose of admitting air and light, but a more important function would be decorating other sceneries and organizing the landscape. The making and structure of empty windows are similar to those of portals. But unlike portals, they have no limit on the shape. Therefore, compared with portals, empty windows

- **石漏窗**

石漏窗采用石材做边框,并用石料仿做木棂格而成。石漏窗结实坚固,适合露天环境,又有漏景的作用,为园林添色不少。本图所示石漏窗是在石板上做出菊瓣纹,比一般的仿木棂格几何纹显得活泼。

Stone Latticed Window

Stone latticed windows are framed with stone and the lattice on them is also imitated with stone. Stone latticed windows are solid and strong, suitable for outdoors. Solid as they are, stone latticed windows are able to leak through scenes on the other side, contributing to the beauty of the garden. The latticed window in the picture has chrysanthemums petal patterns carved on a slate. It appears to be more sprightly and lively than the geometric patterns on other wood-like lattice.

- **苏州拙政园与谁同坐轩的扇面形空窗**

空窗的形制与建筑及环境特点有关,轩、亭、榭多用方形、横长、直长等式样,简洁朴质;廊上的空窗多作连续排列,一般体形不大,式样不同。

Fan-shaped Empty Window of the "With Whom Shall I Sit" Pavilion in Humble Administrator's Garden, Suzhou

The shape of empty window is related to the architecture and the surrounding environment. Empty windows in *Xuan* (studios), *Ting* (pavilions) and *Xie* (pavilions on terrace) are usually square or rectangular, simple and decent. The empty windows on colonnades walls are usually in succession. They are generally small in size and different from each other.

are more flexible in appearance. Enframed sceneries, borrowed sceneries and leaking-through sceneries are mostly realized with the help of empty windows.

- 苏州耦园还砚斋的木雕漏窗

  木雕漏窗采用木材做边框，并用小木条拼接成各种各样的棂格图案。因为木材易于加工制作，木雕漏窗比其他材料制成的漏窗更加轻巧、精致。但不宜在露天环境中使用。

  Wooden Latticed Window in Giving-back-inkstone Studio (*Huanyan Zhai*), in the Garden of Couple's Retreat (*Ouyuan Garden*), Suzhou

  Wooden lattice windows use timber as the frame and small wooden bars to splice grid patterns of all kinds. For the convenience in processing wood, wooden lattice windows are more delicate and light than those made of other materials. However, they are not suitable for outdoors.

- 砖漏窗

  砖漏窗采用异型砖砌成，因所用异型砖不同，漏窗的纹样也不相同。

  Brick Latticed Windows

  Brick latticed windows are laid with special-made bricks. The pattern depends on the shape of the bricks.

- 瓦漏窗

瓦漏窗用瓦砌成，由于瓦的摆法很多，漏窗的纹样也非常多。瓦漏窗的造价较低廉，但装饰效果很好，富有田园风味。

Tile Latticed Windows

Tile latticed windows are lapped with tiles. Since there are many ways to lap the tiles, there are many patterns for tile latticed windows. Tile latticed windows cost relatively less, but the decorative effect is significant—they bring along a countryside flavor.

栏杆

栏杆一般由望柱、寻杖、栏板几部分构成，多用于高台、临水建筑、楼阁、走廊等处，装于两柱之间或窗下，既起到维护安全的作用，又是极好的装饰物。

Balustrades

Generally speaking, balustrades are composed of balusters, handrails, and frieze panels. They are mostly used in high terrace, waterside architectures, storied buildings and colonnades, between two columns or below the windows. They on the one hand ensure the safety and on the other, serve as perfect ornaments.

- 苏州同里退思园水香榭靠背栏杆

靠背栏杆又名"吴王靠""美人靠"，是一种有椅子靠背的栏杆，常用在亭、榭、轩、阁等小型建筑的外围。因其向外探出的靠背弯曲似鹅颈，又名"鹅颈椅"。

Backrest Rail in Water Fragrance Pavilion (*Shuixiang Xie*), Retreat and Reflection Garden, Tongli, Suzhou

Backrest rails, also known as King Wu's chair (*Wuwang Kao*) or beauty's chair, are rails as chairs with backrest. They are usually used in the outer-ring of pavilions, pavilions on terrace, studios. The backrest stretching out curves like the goose's neck, so it is also named as goose's neck chair.

- 北京恭王府花园坐凳栏杆

坐凳栏杆是在廊柱根部安装矮栏，上面架半尺宽或一尺多宽的平板，供人们坐下休息。多用于园林建筑，有时也用于宫殿及府邸宅院中的抄手廊。

Seat Rail in Prince Gong's Mansion Garden, Beijing

Seat rails are short balustrades set near the ground between colonnade columns. They are covered with boards half or one ohi (1/3 meter) wide on the top so that people could sit on them. This style is normally employed in garden architecture, but sometimes it is used in side corridors in palace, mansion, and messuage.

• 北京北海永安桥上的寻杖石栏杆

寻杖指的是栏杆上部的扶手部分。寻杖栏杆多用于檐下栏杆和楼梯两侧的扶手。

Stone Handrails on *Yong'an* Bridge (Bridge of Eternal Peace), in *Beihai* Park, Beijing

Xun Zhang (Handrails) refers to the upper part of the balustrades for the hand to hold. Handrails are usually used on balustrades below eaves and the two sides of the stairs.

• 瓶式栏杆

瓶式栏杆是清代流行的一种西洋式栏杆，栏板采用多根木料旋成的瓶式直棍条，俗称"西洋瓶式栏杆"。

Vase-shaped Balustrade

Vase-shaped balustrade is one popular western style balustrades in the Qing Dynasty (1616-1911). Its frieze panels are usually made of multiple wooden bars in the shape of an upright vase. People call it occident vase-style railing.

• 苏州耦园栏板栏杆

栏板栏杆不设寻杖，只有望柱和柱间的栏板，所以栏板特别突出。有在栏板上作透空雕刻的，也有不作任何装饰、表面平整的，有的还在雕刻的栏板上下加横枋，形式各不相同。

Panel Balustrade in the Garden of Couple's Retreat, Suzhou

Panel balustrades have no handrails, but only balusters and the panels between them. And thus the panels are attached with much importance. Some panels are pierce and carved; some are smooth and without any decorations; some have tie-beams both on the top and at the bottom, and their styles differ.

天花与藻井

天花与藻井是中国古代建筑室内木构顶棚的两种形式。天花源自宋代，可以用来遮挡室内屋顶的梁架结构，保证室内空间的视觉美观。天花可分为井口天花和海墁天花两类。

Compartment Ceilings and Caisson Ceilings

Ceiling (*tian hua*) and caisson (*zao jing*) are two different styles of wooden platfonds in the ancient Chinese architectures. Ceilings date back to the Song Dynasty (960-1279). They could be used to cover the beam on the house top, keeping the indoor space beautiful. Ceilings can be grouped into two kinds: compartment ceilings and flat ceilings.

- 海墁天花

海墁天花表面平坦，没有木条达成的方格，而是用木板做成，或是在较小的房间内架一个完整的框架，上面安木板或糊纸，然后在天花的表面直接进行简单的彩绘，大多是整间屋子的顶部绘画连成一个整体，所以称海墁天花。

Flat Ceiling

Flat ceiling has a smooth surface without wooden compartments. They are made of wooden boards. Sometimes, papers are pasted onto a pre-made frame and then simple drawings are directly painted on the paper. The paintings on the top are a unity and thus this style of ceiling is referred to as *Haiman Tianhua* (sea ceiling).

- 井口天花

井口天花的做法是用木条互相交叉形成若干方格，每块方格中镶进一块天花板，露出的部分通常绘有彩画。井口天花通常用于皇家园林的宫殿大厅中，南方私家园林较少使用。

Compartment Ceilings

Compartment ceiling is a ceiling that has been divided by wooden compartments into decorative panels. The panels are usually painted. Compartment ceilings are usually applied in palace or halls in royal gardens; they are less often used in southern private gardens.

北京故宫乾隆花园古华轩天花

北京故宫乾隆花园古华轩内的天花板用优良楠木做成，方格内为精细的花草图案，雕刻细腻，十分精美，是园林建筑天花装修中最华美精致的一种。

Ceiling of Bower of the Ancient Catalpa, the Forbidden City, Beijing

The ceiling of Bower of the Ancient Catalpa in the Forbidden City is made of fine Phoebe zhennan. The compartments are carved with exquisite flowers. It is one of the most gorgeous ceilings in garden architectures.

　　藻井位于室内屋顶正中央最重要的部位，呈穹隆状，大多由斗拱层层烘托而成，也有用木板制作的较为简单的藻井。藻井常用于宫殿、寺庙的宝座上部屋顶和神佛龛位上屋顶部以及皇家园林中大殿的室内顶部装饰，南方私家园林等建

　　Caisson takes the most important position right in the middle of the roof inside. It is dome-shaped, usually built with layers of brackets (*dougong*). In some cases, relatively simple caissons are made of boards. Caissons are often used at the center of the roof above the thrones, seats and religious figures in palaces or

杭州西湖郭庄两宜轩方形藻井

方形藻井又叫斗四藻井，是一种在早期较为常见的藻井形式，表面呈四方形，形式较为简洁。有的是由两层方形的井口呈45°角错置，也有的在方形井的中心装饰花形，在早期的石窟建筑中则有仿木的石制方形藻井。

Square caisson in *Liangyi Xuan* (Belvedere of Both Good), Guozhuang in the West Lake, Hangzhou

Square caisson (*fangxing zaojing*) is also known as *Dou Si Zaojing*, which is a relatively common seen style for caissons in earlier times. It is square and simple. Some have two squares overlapping with each other at an angle of 45° at the equatorial plane; some are decorated with flower patterns in the center. Stone-made wooden-like square caissons are seen in early grottoes.

筑中则很少用。藻井的形制有四方形、圆形、六角形和八角形等，也有制作复杂的将几种形状融为一体的藻井，层层叠落、精美华丽。

temples, or in major architectures in royal gardens. They are rarely used in southern private gardens. Common shapes of caisson include square, circle, hexagon, and octagon. Some sophisticated designs will combine several shapes together, offering a gorgeous view of cascading structure.

- 北京颐和园廓如亭八角藻井

  廓如亭位于颐和园昆明湖十七孔桥的东端，是园内最大的一座亭子，也是我国现存亭类建筑中比较大的一座。亭内的藻井造型突出，装饰精美，气势壮观，堪称藻井中之精品。

  An Octagonal Caisson in the Spacious Pavilion (*Kuoru Ting*) in the Summer Palace, Beijing

  The Spacious Pavilion, the largest pavilion in the Summer Palace, lies to the east of the Seventeen-arch Bridge over *Kunming* Lake. It also ranks high in all existing large pavilion style architectures in China. The caisson of the pavilion is eye-catching with its exquisite decorations and imposing splendor. It is an elaborate work in caisson.

- 山西晋祠难老泉亭藻井

  难老泉亭是一座八角攒尖顶建筑，亭内的藻井别具特色，形状近似圆形，从中间的井心向外呈放射状，四周和中心都有向下的垂花柱头，更显装饰美感。在藻井的最外层安置斗拱。整个藻井以突出木构件的本色为主要特点，做工简洁、古朴淡雅。

  A caisson in Never Aging Spring Pavilion in Jin Ancestral Temple in Shanxi Province

  Never Aging Spring Pavilion is an octagonal-pavilion-style architecture. Its caisson is unique—it is almost round, and radically symmetrical. There are hanging flower posts around and in the center for decoration. Brackets are set in the outermost part of the caisson. With succinct workmanship and unsophisticated elegance, the caisson is characterized with the natural wood color.

- 北京故宫御花园千秋亭圆形藻井

  A Round Caisson in the Pavilion of One Thousand Autumns (*Qianqiu Ting*) in the Imperial Garden at the Forbidden City, Beijing

铺地

铺地，古代叫"墁地"。园林铺地所用材料有方砖、侧砖、条石、卵石、块石、石板以及碎砖瓦、瓷片等多种，所以园林墁地的花样最多，色彩丰富，风格精巧。

Paving

Paving (*pudi*) is known as *mandi* in ancient times. Garden paving materials include square bricks, bricks-on-edge, strip stones, pebble stones, stone blocks, slates, broken bricks, tiles and porcelain. These various materials enable diversified patterns, rich colors and exquisite styles.

- 苏州留园砖铺地

纯粹的砖铺地一般是用作建筑周围或庭院之间的甬路，用于日常行走，而装饰性相对较弱。

Brick Paving in the Lingering Garden, Suzhou

Brick paving is commonly used for paths around architectures or between courtyards. It offers access, and is less decorated.

- 北京北海鹅卵石铺地

用卵石铺路面，也是一种既经济又讨巧的做法。卵石是天然形成的，虽然大小、形状、颜色不一，但表面都非常光洁圆滑，不会有尖棱硬角，用来铺地，走起来也很舒适。

Pebble Stone Paving in *Beihai* Park, Beijing

Paving the path with pebble stones is both economical and effective. These naturally-formed pebbles, although different in size, shape and color, are all smooth in the surface. When used for paving, the pebbles are comfortable to walk on.

红瓦片
red tiles

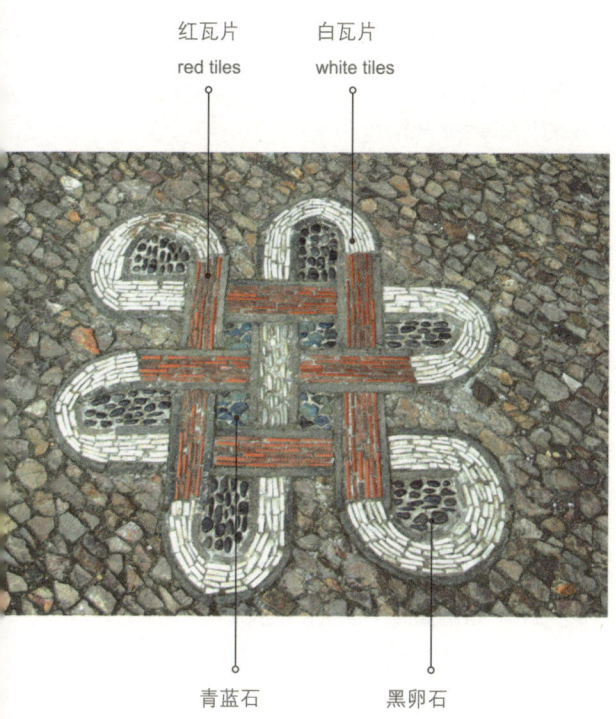

白瓦片
white tiles

青蓝石
blue stones

黑卵石
black pebbles

- **苏州网师园盘长纹铺地**

  同时用砖、瓦、石等几种材料来铺设地面，可以铺设出非常清晰、有趣、富有装饰性的纹样。

  Chinese Knot Pattern Paving in the Master-of-nets Garden, Suzhou

  Application of several materials including bricks, tiles and stones can create patterns clear, interesting and decorative.

- **方砖卵石嵌花路**

  北方宫苑中常用的铺地方法。一般是道路中间铺方砖，方砖两侧铺卵石带，并将卵石拼成寿字、如意、铜钱、扇形、海棠等图案，再以各色卵石填心。

  Brick and Pebble Paving

  This is a common paving in northern palaces. Square bricks are set in the middle of the path, with pebbles paved on two sides. The pebbles outline the character *Shou* (longevity), or patters like *Ruyi* (as you wish), coins, fans and crabapples and then these patterns are filled with pebbles of different colors.

中国园林中的建筑
Architectures in Chinese Gardens

吉祥图案铺地
Auspicious Patterns Paving

凤纹铺地

凤是中国古代传说中的神鸟，被尊为鸟中之王，是祥瑞的象征。园林中以凤纹铺地，取祥瑞之义。

Phoenix Pattern Paving

Phoenix is the sacred bird in ancient Chinese legends. It is revered as the king of all birds and it is considered as a good omen. The phoenix pattern on garden paths is a wish for good luck.

- 苏州留园凤纹铺地
Paving with Phoenix Patter in the Lingering Garden, Suzhou

鹤纹铺地

鹤在民间被视为仙禽，鹤纹有长寿和官居高位的寓意。用作铺地图案，如鹤与鹿在一起，寓意"六合同春"；若一鹤为单腿独立之姿，便是"官居一品"。

Crane Pattern Paving

Among Chinese folk, crane is seen as the immortal bird, symbolizing longevity and high office. In paving patterns, when crane appears together with deer, it stands for thriving prosperity. If it is a crane standing on one foot, it stands for the highest office.

- 苏州网师园鹤纹铺地
Paving with Crane Pattern in the Master-of-nets Garden

鱼纹铺地

民间的鱼纹种类很多，构图也不相同，有多种吉祥寓意。若是金鱼纹，谐音"金玉"，寓意"金玉满堂"；若是鲤鱼纹，则是"连年有余"，含有富裕有余的意义。

Fish Pattern Paving
There are many sorts of fish patterns in the Chinese folklore. Different patterns stand for different auspicious meanings. A gold fish (*jinyu*) sounds like gold and jade (*jinyu*) in Chinese. And thus a gold fish pattern implies prosperity. A carp pattern stands for having more than need every year (*liannian youyu*), implying abundance and richness.

• 苏州留园金鱼纹铺地
Paving with Gold Fish Pattern in the Lingering Garden, Suzhou

鹿纹铺地

鹿在中国自古以来就被当作神兽，象征祥瑞，又因"鹿"与"禄"谐音，鹿纹又代表官俸和财富。鹿的造型多取鹿回头状。

Deer Pattern Paving
Deer has been looked onto as the divine animal since the ancient times in China. It is an auspicious sign. And since "deer" (*lu*) is pronounced in the same way as "salary" (*lu*) in Chinese, the deer pattern also stands for official's salary and wealth. The deer in these patterns are usually looking back.

• 苏州网师园鹿纹铺地
Paving with Deer Patterns in the Master-of-nets Garden

海棠纹铺地

海棠纹源自四瓣的海棠花，形状在方圆之间。铺地用海棠纹为主纹，一是取其好看；二是因为海棠纹工整，易于做四方连续的基础纹；三是海棠纹中有一个"棠"字与"堂"谐音，易与其他纹饰相搭配，表达出更多的吉意。如与玉兰花相组合，则是"玉堂富贵"。

Crabapple Flower Pattern Paving
Crabapple flower pattern originates from crabapple flowers with four petals. Its shape is between round and square. Crabapple flower pattern is usually used for three reasons: firstly, it is good looking; secondly, it is neat and orderly, readily to be applied as consecutive patterns; thirdly, one sound in crabapple (*haitang*) sounds like 堂 (*tang*), which is a good-meaning character that can be put together with other patterns' names for more good wishes. For example, when the crabapple pattern is used with magnolia, they stands for wealth and honor (*yutang fugui*).

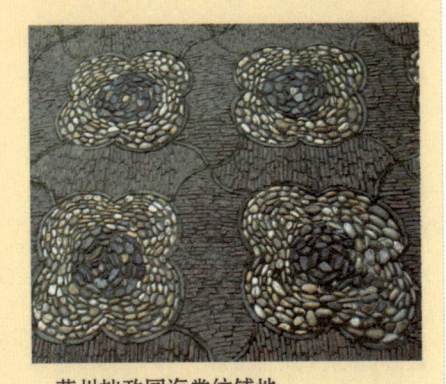

- 苏州拙政园海棠纹铺地

Paving with Crabapple Pattern in the Humble Administrator's Garden, Suzhou

套方纹铺地

套方，是两个菱形压角相叠组成的纹样，又叫"方胜"，相传是西王母所戴的发饰。古时一直作为祥瑞之物，具有辟邪的作用。

Interlocking Squares Pattern Paving (*Taofang*)
Taofang is a pattern composed with a pair of interlocking lozenges, also known as *Fangsheng*. There is a legend that it is the head ornament worn by the Queen Mother of the West. It was used as an object to exorcise evil spirits in the ancient times.

- 套方纹铺地

Paving with Interlocking Square Pattern

寿字纹铺地

寿字纹，寓意健康长寿。在皇家园林中，寿字纹尤为常见。

Shou Character Paving

Shou (longevity) pattern stands for health and longevity. It is especially common seen in royal gardens.

• 苏州网师园寿字纹铺地
Paving with *Shou* Character in the Master-of-nets Garden

彩画

彩画是指建筑木结构上的装饰性彩绘。建筑经过彩绘后，有"画栋雕栏"的精美效果，既可表明建筑主人的高贵身份，还有保护建筑木结构的作用。彩画最早出现在春秋时期的木结构建筑上。明清时期，建筑彩画发展成熟，有和玺彩画、旋子彩画和苏式彩画三大类。

和玺彩画

和玺彩画是清代古建筑中最高等级的彩画。纹样以龙凤为主，用

Polychrome painting

Polychrome paintings are the decorative paintings on the wooden structures of the architecture. These paintings lend charm to the architectures. On the one hand, they are a statement of the nobility of the owner; on the other, they are protection for the wooden parts. Polychrome paintings first exist on timberwork buildings from the Spring and Autumn Period. During the Ming and the Qing dynasties, polychrome paintings on architectures ripened into three major categories: dragons patterns (*hexi caihua*), tangent circle patterns

于皇帝听政、住所等宫殿建筑以及祈天、祭祖等礼制坛庙建筑物上。和玺彩画出现和形成的时间大约在明末清初之际。其特点是枋心用"Σ"线括起，枋心内画龙、凤等各种图案。和玺彩画的设色不同于其他彩画，主体框架线路一律为沥粉贴金做法，不采用墨线。细部纹饰大部分也用沥粉贴金。和玺彩画的纹样与设色也有级别之分，有金龙和玺、龙凤和玺、龙草和玺、莲草和玺。

(*Xuanzi Caihua*), and Suzhou-style patterns (*Sushi Caihua*).

Dragons Patterns

Polychrome painting with dragon patterns is of the highest rank in the Qing Dynasty (1616-1911). With dragons and phoenix as its main patterns, this style of painting is mainly used in the palaces where the emperor administered state affairs or lived, as well as in ceremonial altars and temples for praying to the heaven and worshiping ancestors. It first appeared around the late Ming and the early Qing dynasties. The center of the architraves is enclosed by Σ shapes and patterns of dragons and phoenix are painted in it. Different from other categories of polychrome paintings, the major lines in the painting with dragon patterns are gelled and gilded-no ink lines arc used. Most of the detail patterns are processed in the same way. The patterns are hierarchical, including golden dragons, dragons and phoenix, dragons and grass, lotus flower and grass.

- 龙凤和玺彩画
  Polychrome Painting with Dragon and Phoenix Patterns

• 北京颐和园仁寿殿金龙和玺彩画
Polychrome Painting with Gold Dragon Patterns in the Hall of Benevolence and Longevity, the Summer Palace, Beijing

• 北京故宫体仁阁龙草和玺彩画
Polychrome Painting with Dragon and Grass Patterns in the Belvedere of Embodying Benevolence (*Tiren Ge*), the Forbidden City, Beijing

旋子彩画

旋子彩画是清代官式彩画的另一种形式，等级次于和玺彩画，是用于王府、礼制坛庙或宫廷次要建筑上的彩画。彩画的内容一般是以圆形切线为基本线条组成规则的几何纹样，其中最大的特点是藻头部分的图案，一般是由青绿旋瓣团花组成，规整、淡雅，装饰性很强。

Tangent Circle Patterns

Polychrome painting with tangent circle patterns is another form in official style polychrome paintings, ranking lower than those with dragon patterns. It is used in palaces for princes, ceremonial altars and temples, or subordinate buildings in the palace. The patterns are basically tangentially adjacent circles. A foremost characteristic is the pattern in the intermediate portion of the painted beam. It is usually composed of blue and green flowers with whirling petals. These patterns are with a delicate sense of elegance, regular and strongly decorative.

• 北京颐和园涵虚堂旋子彩画
Polychrome Painting with Tangent Circle Patterns in the Hall of Embracing the Universe (*Hanxu Tang*), the Summer Palace, Beijing

• 北京北海智珠殿牌楼旋子彩画
Polychrome Painting with Tangent Circle Patterns on the Archway of Bright Pearl Hall (*Zhizhu Dian*), *Beihai* Park, Beijing

苏式彩画

苏式彩画源于江南水乡苏州一带，清代时传入宫廷，成为官式彩画中的一个重要品种。苏式彩画构图形式活泼，以花鸟、鱼虫、山水、人物、翎毛为主要题材，色调明快，生活气息比较浓，一般用于园林中的小型建筑以及四合院、垂花门的额枋上。因构图格局及彩画部位不同，可分为包袱苏画、海墁苏画、掐箍头搭包袱苏画等多种形式。

Suzhou-Style Patterns

The Suzhou-style pattern originate from Suzhou. It is introduced into imperial courts in the Qing Dynasty (1616-1911), becoming an important part of official style polychrome paintings. The composition is lively, with birds, fishes, insects, mountains, rivers, feathers and flowers as the main subject. Suzhou-style paintings have vibrant colors and are full of life. The patterns are usually used in smaller garden buildings and on the architraves of courtyard houses and hanging flower gates. According to their composition and location, Suzhou-style patterns can be grouped into categories including wrapper style, flat style, wrapper style with beam-end bands, etc.

• 北京恭王府花园套景包袱彩画
套景包袱以风景为主要题材，或以山水景物与花草植物组合内容。

Polychrome Painting with Landscape Pattern in Wrappers, Prince Gong's Mansion Garden, Beijing

Paintings with landscape pattern in wrappers have different sceneries as their subject. Sometimes there is a combination of landscape and plants.

• 北京颐和园长廊"双燕图"花鸟包袱彩画
包袱是苏式彩画的一种形式，将枋心括起形成一个大的半圆形，犹如一个包袱，在半圆形的轮廓线内绘画各式图案。花鸟包袱彩画多用于南方私家园林，或皇家园林中模仿南方风格的建筑上。

A Pair of Swallows, Polychrome Painting with Flower and Bird Pattern in Wrappers, Long Corridor of the Summer Palace, Beijing

Wrapper is a particular pattern in Suzhou-style paintings. The center of the architraves is enclosed with a large semicircle, which is like a cloth wrapper, and then patterns are painted within this semicircle. Wrapper framed paintings of birds and flowers are mostly used in southern private gardens, or royal garden architectures that imitating the northern styles.

- **人物包袱彩画**

人物包袱彩画多用于私家园林的廊、榭、亭等建筑上。

Polychrome Painting with Figures in Wrappers

This kind of paintings is mostly used in colonnades, pavilion on terrace and pavilions in private gardens.

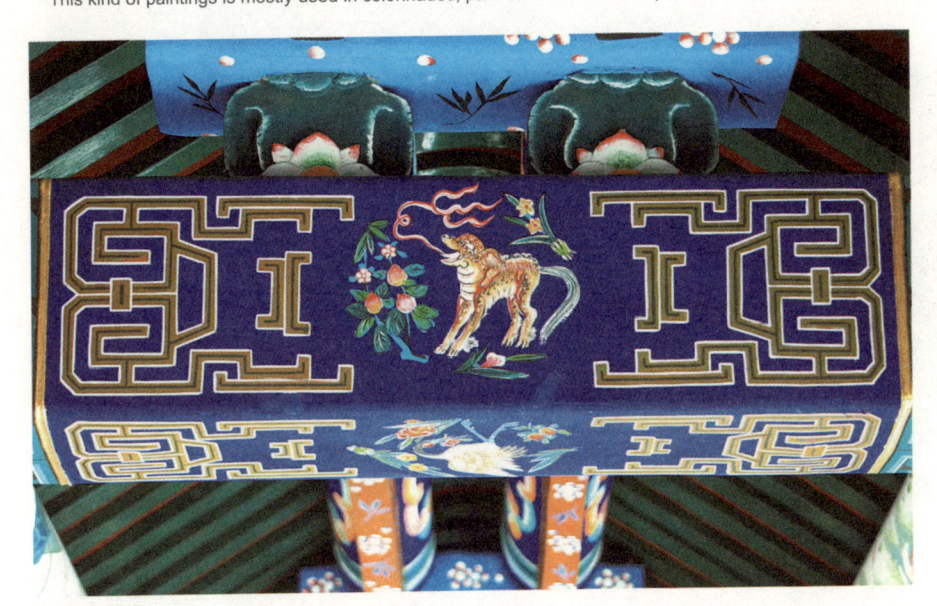

- **海墁苏式彩画**

海墁苏式彩画是一种没有枋心和包袱的彩画，在梁枋上通画一些简单的花纹，如流云或折枝花纹，这是一种级别较低的彩画，多用于建筑的次要部位。

Flat Suzhou-style Painting

Flat Suzhou-style paintings have no wrapper patterns or central portion on the architraves. Simple patterns such as clouds or twigs are drawn on the beams. This style is a relatively lower ranking style, mostly used in less important parts of the architecture.